Walter R. Browne

The Foundations of Mechanics

bremen
university
press

Walter R. Browne

The Foundations of Mechanics

ISBN/EAN: 9783955622794

Auflage: 1

Erscheinungsjahr: 2013

Erscheinungsort: Bremen, Deutschland

@ Bremen-university-press in Access Verlag GmbH, Fahrenheitstr. 1, 28359 Bremen. Alle Rechte beim Verlag und bei den jeweiligen Lizenzgebern.

bremen
university
press

THE

FOUNDATIONS OF MECHANICS

BY

WALTER R. BROWNE, M.A., M. INST. C.E., &c.,

Late Fellow of Trinity College, Cambridge.

———

REPRINTED FROM "THE ENGINEER."

———

PRICE ONE SHILLING.

CHARLES GRIFFIN AND CO.,

EXETER STREET, STRAND, LONDON.

1882.

THE

FOUNDATIONS OF MECHANICS

BY

WALTER R. BROWNE, M.A., M. INST. C.E., &c.,

Late Fellow of Trinity College, Cambridge.

———

REPRINTED FROM "THE ENGINEER."

———

CHARLES GRIFFIN AND CO.,

EXETER STREET, STRAND, LONDON.

———

1882.

FOUNDATIONS OF MECHANICS.

1. It cannot, I believe, be denied that amongst those who have to apply science to practice, and especially therefore amongst engineers, there exists considerable confusion as to the meaning of the fundamental definitions and fundamental principles on which the science of Mechanics is built. These definitions and principles are explained, it is true, in the various works written upon the science; but the explanations are not always as full and as clear as would seem requisite to prevent confusion; nor is it always easy to reconcile, at least at first sight, the definitions and explanations given in one work with those given in another. There would thus appear to be a want of some treatise which shall apply itself specially to the task of setting forth those fundamental definitions and principles in the fullest and clearest manner; using the accounts of them, given in the works above mentioned, as a guide, but supplementing or explaining these where necessary, and taking care to show how they harmonise with the actual facts, as to which, it may be stated at the outset, there is among scientific men little or no dispute. This want was, in fact, strongly insisted upon in a leading article of THE ENGINEER, February 25, 1881; and it is in the hope of supplying it, at least in some measure, that the following pages have been written. They are not addressed to actual beginners in mechanics, but rather to those who have studied the subject in the ordinary way, but who still feel that they need a firmer and surer grasp of the principles, especially in order to be able to apply them with confidence in practice. It will be assumed, therefore, that the reader is familiar with the leading facts and propositions, both of mechanics and of engineering, and free reference will be made to these whenever necessary. For the same reason no attempt will be made to divide the subject sharply into

the three branches of statics, dynamics, and kinematics—a division which is convenient for the purpose of elementary instruction, but is in some ways unfortunate as regards the study of fundamental principles. On the other hand, the treatise is still less intended for advanced students of higher dynamics, whose perfect familiarity with, and agreement in, the symbolic form of the science, renders them comparatively indifferent to the names employed for its elementary conceptions. Their business is to rear and to ornament the building; mine is the humbler one of attempting to give an accurate plan of the foundations—a part of the structure which architects and householders are both somewhat inclined to neglect, but an acquaintance with which is necessary alike for the stability of the building and for the security and comfort of its inhabitants.

2. Before beginning to consider the definitions of mechanics, it will be well to make one or two remarks about definitions in general. It is necessary, in the first place, to draw a distinction, very important but often overlooked, between definitions of terms and definitions of things.* The nature of this distinction is well illustrated by the definitions used in algebra and in Euclid. When we proceed to prove a theorem or solve a problem by algebraical methods—for instance, a problem as to the number of acres in a field which is reaped under certain conditions— we begin by saying, "Let x = the number of acres in the field." We have then defined the term x for the purpose of that particular problem; from henceforward it stands merely as a convenient symbol for the words by which it is defined, and if we take care to preserve its meaning unaltered, we shall solve the problem much more easily and clearly by its aid. And this will not prevent us from defining the same term as something quite different—say the number of gallons in a particular tank—for the purposes of the next problem we may wish to attack. In both cases we are merely defining a term, and have only to take care

* This difference, which is, of course, an old one, is objected to by J. S. Mill—Logic, p. 296—but its reality and utility will be established, I believe, at least as far as mechanics are concerned, by the present treatise.

that we keep the term to its definition. But when Euclid defines a square as a quadrilateral figure, of which all the sides are equal, and all the angles right angles, he is not telling us in what sense he is going to use a particular *word*, but is giving a sufficiently accurate description of a particular *thing*, namely, a geometrical figure of which everybody has a general knowledge, and the exact properties of which it is his purpose to investigate. And all his other definitions will be found to be of a similar character.

3. From the distinction thus drawn several consequences follow. It is evident, for instance, that the same word may properly have several different definitions; different, that is, not merely in the exact words used, but in the conception which those words convey. But of these various definitions one only can be the definition of a thing— omitting the case of synonymous words, such as "race," as to which there is practically very little confusion : the others must only be definitions of terms. Thus, as we have just seen, the letter x, taking it as a word, may be defined as a term in innumerable different ways; but as a thing it can have but one definition, which would be somewhat as follows :—A letter in the English alphabet expressing a particular sound, which sound can of course be only spoken, not written. If a word is used indiscriminately of two things which are not, like the two meanings of "race," entirely different, but yet which cannot be brought under the same definition, confusion is nearly certain to result; and if the practice exists, and has gone too far to be stopped, the only remedy is to banish the word from precise and scientific language altogether. To give a single instance, the word "nature" has been used so widely and loosely, that it would almost certainly be impossible to construct a definition which should cover the whole of its applications ; and accordingly it should never be used where accurate writing is intended.

4. It should, however, be noted that in giving several successive definitions to the same word, care should in general be taken to preserve some connection between them. Thus the connection between the innumerable definitions given to

x in algebraical problems, is that in every case it expresses the unknown quantity which is the subject of inquiry. Similarly in Co-ordinate Geometry, x expresses the co-ordinate of a point as measured not up but across the paper. If in any particular case we were to reverse this, and call x the co-ordinate measured up the paper, we should be extremely likely to get our work into confusion. Similarly, in giving a scientific definition, as a term, to a word which is used in ordinary speech, it is most desirable that the scientific should not be inconsistent with the ordinary acceptation. If it is, confusion is sure to follow, from the natural error of sometimes mixing up the meaning of the scientific term with that of the ordinary word—a fallacy than which none is more common in argumentative writings of all kinds. Thus, if we were to define nature as "the totality of all phenomena"—a definition actually proposed—we should certainly run into confusion when arguing with ordinary people, who recognise the possibility at least of phenomena which are supernatural, or beyond nature.

5. Further, it will be evident that the definition of a term will generally be much more meagre, but at the same time more complete, than the definition of a thing. A term may be used, and used correctly, of a number of things, which perhaps have no property whatever in common, except that to which the term applies ; and in that case the definition can express nothing beyond that property. Thus, if we define an explosive as "a substance having the property, under the influence of heat or impact, of suddenly generating a large quantity of gas," that definition cannot be objected to on the ground that it does not really tell us what an explosive is ; in other words, does not give us tests by which we may recognise an explosive as soon as we see it. The fact is that no such definition is possible. Under the term explosives are combined a large number of chemical substances, which, except as to this particular property, have little or nothing in common ; and as long as we deal with the theory of explosion alone, the definition given above is quite sufficient for our purposes, and indeed the only one, probably, which could be used with satisfaction.

6. Lastly, we may draw from what has been said the

obvious conclusion that there must be some things which cannot be defined. For every definition must be in words, and each of those words may be challenged for its own definition; and if this is persisted in without limit, we must either come round in a circle or consent to an endless retrogression. We must therefore take our stand upon certain simple words, representing things so familiar that they can be no further elucidated; and these must form the elements, out of which all other definitions are made. Thus, if I am asked in the last result to define existence, I simply reply, "Existence is that which I mean when I say that I myself exist;" and I refuse to be driven further than that elementary fact. It may seem a pity that, in all the past ages of learned disputation, no attempt has been made to settle what these elementary things shall be taken to be, and by framing a list of them to establish a common basis of argument; but since no such attempt has been made, each man must frame his own list in the best way that he can.

7. In the above remarks on definitions in general, we have avoided, as far as possible, illustrations taken from mechanics —the science under consideration. Their application to that science will become clear in the course of our investigation, to which we may now proceed.

8. *Definition of Mechanics.*—Since our business is exclusively with definitions and first principles, we must begin by defining the science itself of which we treat. For this purpose we shall adopt the definition of Rankine ("Applied Mechanics," Introduction, Art. 1), which is as follows :— Mechanics is the science of rest, motion, and force.

9. To this definition Whewell ("Mechanics," p. 3) practically adds another clause, defining mechanics as "the science which treats of the motion of bodies—or which treats of forces—so far as they are governed by discoverable laws." And this clause he justifies by the following weighty words: —"In many chemical, electrical, and magnetical phenomena the motions of bodies occur; but in those cases the circumstances and laws of the motion are not considered; if they were, that part of the reasoning would belong to mechanics. It is probable that almost all the phenomena, in the different departments of natural philosophy, consist in the insensibly

small motions of particles; and if we knew the laws of their motions, these sciences would, so far, become branches of mechanics; hence, it is probably only the imperfection of our knowledge which prevents the greater part of natural philosophy from being included in the science of motion."

10. It cannot be denied that the probability here spoken of has become very much greater in the sixty years which have elapsed since these words were written. We already talk familiarly of the mechanical theory of heat; electricity and magnetism are found to follow mechanical laws; and the time can hardly be far distant when all the natural sciences—at least those concerned with inorganic matter—will be recognised as branches of mechanics. But, on the other hand, we have little room for doubt that in the case of all sciences the motions are governed by discoverable, though as yet undiscovered, laws; and therefore the clause which Whewell inserts in his definition would appear an unmeaning restriction. At the same time, the vast future extension of the science which he here anticipates furnishes an additional argument for the need of attaining an exact and accepted terminology with regard to its fundamental conceptions.

11. Accepting Rankine's definition of mechanics, we shall also adopt his names for the three divisions into which, from the very words of the definition, the science naturally falls; namely (1) kinematics, being the science of motions considered apart from forces; (2) statics, being the science of forces considered as producing rest; (3) dynamics, being the science of forces considered as producing motion. At the same time, as mentioned at the outset, we shall not confine ourselves strictly to these divisions in the present treatment of the subject.

12. Whilst agreeing in the matter of the above definitions, Thomson and Tait—"Natural Philosophy," p. 1—prefer a different order in the names. What we have called mechanics they style dynamics; and they divide it into the two parts of statics and kinetics, the latter taking the place of our dynamics; while kinematics remains to represent the science of pure motion. Mechanics they prefer to use as signifying the practical science of constructing machinery,

which appears to have been the original meaning of the word. But the Greek word from which the term mechanics is directly derived has a very much wider meaning than our word machine, and may fairly be applied to any arrangement or apparatus by which changes are, or might be, effected in the position, constitution, or circumstances of bodies; to anything, in fact, by which force is brought to bear. Hence the term mechanics seems sufficiently well adapted to express the science of forces and motions. Moreover, the distinction of statics and dynamics, and the use of mechanics and mechanical, in speaking of the general science, have been too long and too widely adopted, in England, France, and Germany, to be easily shaken, even if philology imperatively claimed that they should be reformed. As a matter of fact, the nomenclature of Thomson and Tait does not seem to have met with general acceptance.

13. We define our subject then as the science of rest, motion, and force. And it is obvious that we may next be asked to define what we mean by these three terms.

14. *Definition of Motion.*—It may fairly be held that motion—including rest, which is merely the negation of motion—is one of those elementary facts to which no definition can add anything in clearness ; and accordingly Thomson and Tait give no definition of it, while Whewell contents himself with observing that the idea " is obtained or suggested, as we observe the changes of situation in things around us." Rankine explains motion as " the relation between two bodies when the straight line joining them changes in length, or in direction, or both ;" but this seems rather a result of motion than a definition of it. Accepting the idea of space as a final one, for our purposes at least, motion proper, or absolute motion, would mean a continuous change of place from one fixed point in space to another ; but the existence of such motion can be recognised by us only through a change of position in the thing moving, with regard to some other thing assumed to be fixed. If there were anything which we knew to be absolutely fixed in space, we might perceive absolute motion by change of place with reference to that thing. But as we know of no

such thing, it follows that all motion, as tested and measured by us, must be relative—must relate, that is, to something which we assume to be fixed for the moment. Hence the same thing may often be properly said to be at rest and in motion at the same time; for it may be at rest with regard to one thing, and in motion with regard to another. Thus, take the very homely instance of a man punting his barge up a river by leaning against a pole which rests on the bottom, and by advancing his feet successively on the deck as if walking. Such a man is in motion relatively to the barge; he is also in motion—but in a different manner—relatively to the current; he is at rest relatively to the part of the earth immediately under his feet; he is in motion relatively to the polar axis of the earth and to the sun; whilst it is easy to imagine a proper motion of the whole solar system such that he would be absolutely at rest in space.

15. Motion, then, as regards the observer, is the change of place of the moving thing with reference to some other thing, supposed fixed; as regards the moving thing, it is a condition in which it is changing its place with regard to some other thing. And from these facts—that motion is relative, and that it is a condition—we see at once the error of those who assume that motion is a *thing*, an object with a separate existence of its own, which can be measured and handed from one to another, and is by nature indestructible. Motion can no more be dissociated from a thing moved, than pain from a person pained, or decay from a thing decaying. That this is so is shown in common speech, since we can only speak of a thing being in motion, or the motion of a thing. We cannot speak of a motion without reference to some thing any more than of a pain without reference to some body. Hence it is clearly a mistake in language to speak of a quantity of motion, since quantity implies the measurement of some definite thing; and accordingly the old term "quantity of motion," though only employed in a technical sense, has been properly replaced by "momentum." We cannot speak of the quantity of a pain: we speak of its intensity; and so we might speak of the intensity of a motion, were it not that this has a specific name of its own, namely, velocity.

16. For the same reason, in using accurate language, we must not say that motion is *transferred* from one body to another. If A, finding himself in a passion with B, uses injurious terms which put B in a passion also, we cannot say that A has transferred his passion to B, even if his own anger has been somewhat relieved by the explosion. Similarly, if a body A strikes against a body B, and puts it in motion whilst stopped itself, we must not say, speaking scientifically, that A has transferred its motion to B. It has only put B into a like state—with reference to some assumed fixed point, be it remembered—to that in which it had previously been itself. This may appear an unnecessary strictness of language ; but there seems reason to think that the frequent use of the word transference, as applied to motion, is one of the causes which have led men to consider motion as a thing instead of merely a condition.

17. The above does not, of course, imply that there is no such condition as absolute motion, or that practically there are any bodies which are not in a state of absolute motion; but only that the direction and the intensity of absolute motion are for us impossible alike to recognise and to measure, and therefore that no conclusions can properly be drawn from it.

18. *Definition of Force.*—We have now defined mechanics as the science of force and motion, and we have explained, in the impossibility of defining, the sense of the word motion. We must next define, or explain, the sense of the word force. The definition we shall give is a very brief one, and is this: —*A force is a cause of motion.*

19. This definition is substantially that given by most writers on mechanics; but by those who approve it, it is usually expressed in an expanded form, while it is disapproved altogether by others. It is therefore necessary to defend it on both these sides.

20. The expanded definition is that given by Newton, "Principia," Def. v.,* and is nowhere put more clearly than

* The literal translation of Newton's wording is :—Impressed force is an action exercised on a body, tending to change its state either of rest or uniform motion in one direction."

in the "Course of Mathematics" by the present Bishop of Carlisle, where it appears as follows:—"Force is any cause which changes, or tends to change, a body's state of rest or motion." The object of this expansion is of course to bring out the fact that forces may act in many cases without causing motion, *e.g.*, when a weight rests upon a table. But the fact in such cases is not that force, as a cause of motion, ceases to operate, but that its operation is exactly balanced by an opposite or counteracting force. Now we are perfectly familiar with the idea that a cause may exist, and yet may be prevented by an opposing cause from producing the whole or any part of its ordinary effect. For instance, we certainly hold that the attraction of the earth is the cause of the fall of bodies, although from opposing causes many bodies remain suspended above it, or even rise. Again, we say that the application of fire to gunpowder causes an explosion; although we know that if the powder be wet, the explosion will not follow. Or, again, we say that radiant heat is a cause of the sensation of warmth in my hand; although in any given case, *e.g.*, if the hand is numbed, or at about the same temperature as the source of heat, that sensation will not be experienced. That force, therefore, although the cause of motion, may not in all cases produce actual motion, is not really a separate part of the definition, but is deducible from the fact that force is a cause. The deduction will properly take the following form:—A force, being a cause, will produce motion, unless it is prevented from doing so by a counteracting force. In practice, the cases where it is so counteracted are, of course, numerous and important; and the investigation of these forms the science of statics.

21. Turning next to those who disapprove of the definition altogether, we cannot select better exponents of this view than Thomson and Tait, who observe, "The idea of force, in point of fact, is a direct object of sense; probably of all our senses, and certainly of the 'muscular sense.'" Hence they regard force as an ultimate fact of consciousness, like warmth, sweetness, or any other sensation, and refuse to define it. Now, I fully concur in the view that force is an

ultimate fact of consciousness,* though it may be doubtful whether it is properly classified as an object of sense; but I justify the use of the definition by appealing to the distinction drawn in the introduction between definitions of things and definitions of terms. As a *thing*, force is indefinable. We can only say, "Force is that exertion of which I am conscious when I pull or push or lift, or do anything of which I say that it requires force." But the *term* force, as used in mathematics, may be defined, and is defined with advantage in the words here given. The advantage is two-fold. In the first place, by means of this definition, we can exclude from mechanics a number of questions which might otherwise be asked, and would have to be answered, although the answer has nothing to do with the science. For instance, we get rid of the complaint which is sometimes made that writers on mechanics do not, after all, tell us what force is, in its essence and real nature. For answer, we only reply that we define force as a *term*, and that, as pointed out in the introduction, the definition of a term does not imply or require any other statement whatever as to the various objects in nature to which that term may be applied. For instance, we do not assert that what we call gravitation, and what we call magnetism, have any connection whatever, except the fact that they both produce motion, and are therefore forces; while still less do we assert that we may not, in the further progress of this or other sciences, be able to trace out some further connection between them.

* This view of force is also taken by Mr. Herbert Spencer, in his "First Principles;" but I altogether dissent from his further statement, that the idea is given by the impressions—of resistance, &c.—made on us by external objects. Let the reader rest his hand on the table with a book upon it, and then let him begin slowly to exert and to increase force until the book is just lifted. The sensation produced by the book upon the hand does not alter. What supervenes is a sensation, if it can be called so, in the muscles of the arm, accompanied by an involuntary inspiration, and general feeling of tension throughout the body. The idea of force is given by what we exert ourselves, not what other bodies exert upon us. Of course, it does not follow but that there must be a resisting body to give us the feeling. The difference is like that between an idea and a thought—we may not be able to think without an idea to think of; but idea and thought are not the same things.

22. In the second place, we are able to show that certain principles, which must otherwise be laid down as separate and independent axioms, flow in reality from the definition, either directly, or by the aid of general principles, which do not belong to one science only, but are common to many. This is of great importance, inasmuch as the structure is thus shown to rest not on so many separate props, but on a stable and connected wall. One instance of this we have given already, in showing that the definition of force as a cause leads at once to the principle that forces may be counteracted by other forces, and that an expansion of the definition to include this fact is not therefore necessary. And we will now give two others, which are, perhaps, still more important.

23. It is sufficiently clear that all causes external to myself can only be known to me by their effects, direct or indirect, upon myself. For instance, if I am eating my dinner in a room in London, I may be receiving from external objects impressions of sight, hearing, smell, taste, and touch, all of which have no existence for my neighbour, only a few feet distant, but on the other side of a thick party-wall. The causes are there, in his immediate neighbourhood; but he knows nothing of them, because they produce no effect upon him. And if causes are only known by their effects, it follows that they can only be estimated or measured by their effects. In such measurement, however, we must take care to begin by examining the effects of the causes on one and the same object—since different objects may produce a specific difference in the effect—and also during the same times. Now apply this to the case of force, defined as the cause of motion. Then it follows that we can properly measure a force—or, in other words, we can properly compare the magnitudes of two forces—only by comparing the intensity of the effects, that is the intensity of the motions, which they produce in one and the same object. But the intensity of motion is velocity, and hence we have the principle, that when forces are applied to the same object—or to objects in all respects equal—they are measured by the velocities which they generate in a given time.

24. The second principle referred to is derived from our definition by the help of an axiom, which is perhaps the widest and most important generalisation which science has yet been able to make in the domain of nature. It is so vast, nay universal in its scope, that it includes within itself, or within its converse—as we shall hope to show at another time — many other principles which are generally stated independently, as generalisations of the first rank in width and importance. Indeed the universal principle has been so far merged in these, which are really particular cases of it, that it has hardly as yet received a distinct standing and name of its own. It will, perhaps, be best understood under the name of the Principle of Conservation. And it will, perhaps, be most clearly, if somewhat metaphorically, ex-pressed by two simple words—Effects live. By this is meant that the effect of any cause does not die away or cease when the cause is withdrawn, any more than the life of the offspring dies or ceases when it is separated from the mother; nay, more, it will not cease at all, but will continue to live by its own vitality, as it were, unless and until it is violently put an end to by some other action of the opposite character. In a word, an effect does not cease ; it is only destroyed. And even when destroyed it is not as though it had never been; for its destruction in itself produces an effect, and in some way an equivalent effect, on the agent which has destroyed it; and by the same law this effect also lives, unless and until it likewise is destroyed by some third agent, to which in turn it also communicates an equivalent effect; and so the generation is continued for ever.

25. The proof of this great generalisation, like that of all other generalisations, lies mainly in the fact that the evidence in its favour is continually augmenting, while that against it is continually diminishing, as the progress of science reveals to us more and more of the workings of the universe. That it is true to some extent is shown by such every-day facts as that a stone continues to fly after it has left the hand, that waves continue to roll after the wind has dropped, that the horseshoe continues to glow after it has been withdrawn from the fire, and so forth. On the other hand the apparent exceptions—*i.e.*, the cases in which effects seem to die

away altogether, after a longer or shorter interval—are so many that it is not to be wondered at if for many ages the principle failed to impress itself on the human mind. But the progress of modern science has shown so many of the exceptions to be apparent only, not real, and has at the same time brought to light so many additional instances of the rule, that the current of thought has changed; and the danger is now lest men should follow the rule too blindly and implicitly, and extend it to regions where it has not been shown to hold. Among the exceptions explained, may be mentioned as a signal instance the discovery that where mechanical work disappears it is always converted into some equivalent, such as heat; and among the new illustrations of the rule, the noblest is, perhaps, that furnished by astronomy, which teaches us that the majestic sweep of the planets through space is due to their having once for all been set in motion by something beyond themselves, somewhere, somewhen, somehow.

26. Without dwelling further on the proof of this great principle, we may proceed to apply it to our definition of force. If force is the cause of motion, and if effects live, then the particular effect called motion lives. In other words, a body once acted on by a force will retain precisely that intensity and direction of motion which the force has left it with, unless and until some other force supervenes to cause a change. In more definite language, *a body, under the action of no external force, will remain at rest, or move uniformly in a straight line.*

27. This is Newton's first law of motion, which is usually stated as an independent principle of dynamics, but which is thus seen to flow from the definition of Force by the simple application of the universal principle of Conservation. It is a well-known fact that the want of appreciation of this principle delayed for ages the progress of dynamics; because men thought themselves bound to look for the force which kept up the motion of a body, instead of simply looking for the force that had started it. This fact may at least teach us not to fall into the opposite error of regarding the principle of conservation as a necessary truth. There is no *à priori* reason to be given why effects should not die

away of themselves, either at once or by degrees. Looking to the continual instances of decay around us, this seems to me even now the easier and more natural supposition; and I only accept the opposite because the facts of the universe force it upon me.

28. There are other fundamental principles, besides the first law of motion, which may be deduced in the same way from the definition; but it will be better to postpone these to a later stage.

29. *Cause.*—We must not leave the definition of force without saying something about the other essential word which it contains, namely, cause. To discuss this word thoroughly in the light, or shall I rather say in the obscurity, of all that has been written upon it, would fill a volume, and would take us off the solid ground of science into the confused and misty limbo of metaphysics. But there are some few at the present day who appear to disapprove of the word altogether, and to imagine that it may and should be done away with. Recognising the fact that external causes are only known to us through their effects, they apparently infer that we do not know them at all, and have no right or need to suppose their existence. Thus in mechanics they would eliminate force altogether, and pursue the science with the conceptions of motion alone. In answer to this it might be thought sufficient to appeal to the universal acceptation of the reality of causation; to the effect that even sceptical writers like J. S. Mill and Herbert Spencer found their theories of things upon it, and regard the existence of an effect without a cause as absolutely impossible either to credit or to conceive. I am not myself, however, disposed to adopt this position, and would rather urge upon these despisers of causes the following more practical considerations.

30. (*a*) The history of science is the history of the discovery of causes; her advances have been made on the single plan of studying events with a view to determine the causes of them. Thus the objectors are urging her to forsake a road which is thoroughly explored, and has already led to splendid discoveries, in order to follow one whose

course and end are alike unknown. To take the instance of mechanics, they would abolish what is called dynamics, and merge it altogether in kinematics. Let them then produce a new edition of Newton's "Principia," or Thomson and Tait's "Mechanical Philosophy," in which all the results shall be fully and satisfactorily proved, without the word or the conception "force" being anywhere used in the proofs. It will then be time for them to claim that these works should be henceforth thrown aside as superannuated lumber.

31. (*b*) It is true that external causes are only known to me by their effects; but these are not the only causes. I am myself a cause; and here it is the cause which is known directly, and the effects which are known indirectly. Thus I have already set forth the view, shared by men so different as Whewell, Thomson and Tait, and Herbert Spencer, that we have an immediate and ultimate conception of force, as causing those motions which we produce; and have therefore exactly the same reason to assume its reality as to assume the reality of thought, of sensation, of space, or of anything else which forms a primary fact of consciousness.

32. (*c*) As we have already seen, motion cannot be looked upon as a thing in itself; it must always be the motion of something. Now that something cannot always be ourselves, otherwise we could treat of no motions except our own. Hence it must be something external to ourselves, and therefore known only by its effects upon us. But the motion is not the motion of the effect; the effect—the sensation or impression produced upon us by a moving object—has no motion, or a motion wholly different from that of the object. Hence there must be something which moves, and this something is not an effect upon us, and yet is indissolubly connected with some effect upon us. Now this connection is precisely what we express by the word causation. Hence if there is external motion, then there is something which moves, and that something is a cause. Therefore those who admit that there is such a thing as external motion, must admit that there are such things as causes; and, if so, those causes must be worth investigation, and science cannot be complete unless it investigates them. Moreover, if there are causes for impressions in general, it becomes probable that

there are causes for motion also; and this probability becomes almost certainty when we remember, first, that we are directly conscious of ourselves as causing motion by exerting force; and, secondly, that science strengthens daily the proof that all impressions made upon us are due to motions of some kind or other.

33. *Definition of Matter.*—The definitions of force and motion do not enable us to proceed at once to the first principles of the subject, for we have seen that motion implies something which moves, and I may be asked to define this something. The general name given to this something is *matter*, and we may therefore begin by saying that matter is that which is moved by force; but this is not really a definition, although given as such by some writers; it is merely a substitution of the name "matter" for the general word "something." We go on to inquire whether any definition, whether as a term or a thing, can be given of matter.

34. The writers on mechanics treat this question in ways which vary considerably. Some, as Moseley and Maxwell, attempt no account of it, and simply proceed to speak of matter, or of bodies, as things with which their readers are familiar. Others, as Whewell, define body or matter as "the most general name which we give to everything which is the object of our senses." This explanation, though suited perhaps to metaphysics, is not of the character required for dynamics. Rankine—" Applied Mechanics, Introduction " —defines matter as "that which fills space;" but he omits to state whether this means that which really fills space, or that which apparently fills space—a very important difference. Moreover this definition touches a much-vexed question, which there is no necessity to solve for the purposes of mechanics. Thomson and Tait—ch. 2—observe: "We cannot, of course give a definition of *matter* which will satisfy the metaphysician; but the naturalist may be content to know matter as *that which can be perceived by the senses,* or as *that which can be acted upon by, or can exert, force.*" The former of these definitions, like Whewell's, is a metaphysical rather than a physical one; and moreover, seems

c

open to the objection that what is directly perceived by the senses—*e.g.*, warmth, light, sweetness, pain—is not matter, but an effect of some condition of matter. The latter seems at first sight the same as that suggested in the last paragraph; but we immediately see that it includes something else, and that something of the greatest importance; for it says that matter can not only be acted upon, but can act; can not only be moved, but can also cause motion. It does not, of course, imply that there is one kind of matter that acts and another that is acted upon; one kind which causes motion, and another which is moved. The whole of mechanics is built on the assumption—explicitly stated, in fact, in Newton's third law of motion—that matter both acts and is acted upon. Hence, looking forward to the fundamental principles of the science and the mode in which they are proved, I propose to formulate the definition of matter, *as a term in mechanics*, as follows :—*Matter consists of a collection of centres of force distributed in space, and acting upon each other according to laws, which do not vary with time, but do vary with distance.**

35. This definition is of matter in general; but in practice we are always treating of some definite portions of matter, and we require names to express these portions, according to their size and other properties. The names usually employed are the following, beginning with the most elementary.

---

* It may be well to say something of the terms which occur in this definition. "Collection," "distributed," "space," "laws," are all ordinary and well-understood words, which need not detain us. The laws in this case are laws the mathematical expression of which is not a function of time, so that, other things remaining the same, the force acting is always the same from one moment to another. On the other hand, the mathematical expression of these laws is a function of the distance between the two centres considered, so that as this distance varies the force acting also varies. Lastly, the conception of a "centre of force" is one which becomes familiar to every student of analytical dynamics. It is that of a point in space movable or fixed, from which, as a centre, force is exerted in all directions upon all other points which have the capacity of being acted upon by it; in contradistinction to a point which does not act on others at all, or only on points in certain positions, *e.g.*, along a particular line, or on a particular surface.

36. (1) The centre of force itself is called an ultimate atom, or a physical point. Like a geometrical point, it has no assignable parts or magnitude, and cannot therefore be compressed, extended, or divided. It is, in fact, a geometrical point, conceived as having also the properties of exerting and receiving force, and of being movable through space, whilst retaining its constitution unaltered. The word "point" is, perhaps, the simplest and shortest which can be applied to it, but in this treatise I shall continue, for greater clearness, to use the word "centre."

37. (2) A collection of points or centres, acting on each other, and so intimately and closely bound up together that no known process or natural force can separate them, is called an atom or a molecule. I shall here follow Clerk-Maxwell in using the latter term.

38. (3) A collection of points, simply considered as so small that for the purpose of any particular investigation, or for those of elementary mechanics in general, it may be considered as concentrated in a single central point, is called a particle. This word is used merely to imply that all questions of size, rotation, constitution, &c., are for the present left out of account.

39. (4) A collection of points of any size or shape whatever, which for the purpose of any investigation is treated together as a whole, is called a body.

40. On the foregoing definitions the following remarks may be made.

41. (a) The definition of matter, as stated, is only its definition as a term of mechanics, and only relates to it as it is concerned with force. It does not assert or deny that matter may have other properties; e.g., the properties which distinguish the different chemical elements *may be* special properties of so many different kinds of matter. If, however, chemical properties, &c., should eventually be resolved into manifestations of force, this distinction would cease; and thus, what is now the definition of matter as a *term* may eventually prove to be also its sufficient definition as a *thing*.

42. (b) The definition covers all those properties usually

classed as general properties of matter. Extension on this view is recognised as a property, not of the centres of force themselves, but of the space in which they are distributed; as it is of all space. Hardness, colour, temperature, penetrability, &c., are all now recognised as properties depending merely upon force.

43. (c) At the same time the definition does not absolutely commit us to the statement that the centres of force are in the strict sense infinitely small. All I insist is that no assignable magnitude can be allowed to them. If anyone prefers the conception of a magnitude which, though less than any assignable magnitude, is yet not infinitely small, he is not precluded from it.

44. (d) The definition does not preclude the existence of different *kinds* of matter. Thus there may be two kinds— A and B—such that all centres belonging to kind A act on each other according to one and the same law, but act on those of kind B according to a different law, or even do not act on them at all. For example, there are certain facts which appear to show that the attraction of gravitation does not exist between the molecules of the ether and the molecules of a crystal; on the other hand, the fact that the molecules of the crystal are thrown into agitation by the radiant heat of the sun— which is an undulation in the ether—seems to prove that *some* action takes place between the two sets of molecules. Assuming this, we should say that the molecules of the earth and of the ether belonged to different kinds of matter. But elementary mechanics does not consider the motions of the ether, and therefore for the purposes of this treatise we may consider all matter as of one kind; in other words, that the centres of force are all alike, and all act on each other by like laws.

45. (e) As already mentioned, some writers, such as Maxwell, avoid the use of the word "matter" altogether, and prefer to speak of a "body," in all cases understanding thereby a portion of matter of any size and form; and in the article of THE ENGINEER already referred to, and also in a paper by Dr. Lodge—*Phil. Mag.*, 1879—this practice is formally approved and adopted. For many purposes there is no harm in this course; but when we are trying to

give a rational account of the principles of the science, with a view to their application in practice, it appears to me open to grave objection. For the general result of the application of force to a finite body is to produce three different effects: —(1) a translation of the whole body in some direction; (2) a rotation of the whole body about some axis; (3) a strain or internal displacement of the different parts of the body. Now to consider all these three effects together forms the office of the highest and most difficult branch of mechanics; and therefore for elementary purposes it is essential to consider them separately, beginning with the simplest—that of translation. But if we take as our typical case that of a finite body of any size, we can only treat the case of translation separately by making two assumptions. First, to get rid of rotation, we must assume that the direction of the force passes exactly through the centre of gravity of the body. This assumption is realised, at least approximately, in many cases of practice, but by no means in all. Secondly, to get rid of internal strains, we must assume that whatever be the external force, the body preserves the arrangement of its parts absolutely unaltered, which is expressed by saying that the body is rigid. This is an assumption which is a very large one indeed, inasmuch as it amounts to saying that the internal forces of the body are infinite; and it is needless to add that it is *never* realised in practice. Thus, the extension of a steel bar becomes a very visible quantity long before it breaks; or again, the commonest of all engineering problems, that of the strength of a beam or girder, cannot be approached at all without assuming that it *has* changed its form under the load. Now, to make this vast and radically false assumption the very basis of our dealings with the science, appears to me a course fraught with danger; it is almost certain to produce confusion, if not in the mind of the teacher, at least in the minds of the taught. On the other plan, by beginning with the simple element, or centre of force, and considering first the action of one centre of force on another, we get rid of the need of both assumptions; for the force must pass through the centre of the body acted on, since it is nothing but a centre; and there can be no strain of its internal parts, since it has no internal parts. In this arrangement, I only follow what is surely an accepted rule in the teaching of science, namely, to begin with a single case, and with the simplest elements, and thence to rise gradually to compounds, and to complicated arrangements.

46. (*f*) What has just been said is not, of course, to be read as objecting to the legitimate use of the hypothesis of rigidity. When the principles and lower branches of the subject have been mastered, it is perfectly open to make this approximate hypothesis for the purpose of attacking some of the higher problems, which are more easily mastered thereby; and, in fact, this process forms a separate branch of the subject, under the name of rigid dynamics. But the first and lowest branch of analytical dynamics is called dynamics of a particle; and this branch goes throughout on practically the same assumption as that here advocated—namely, that the body considered has no parts to produce rotation or strain. The only objection to the use of the word "particle," instead of "centre," or "physical point," is that it leaves out of sight the fact that the body considered, besides being acted upon, is itself in all cases a source of action; and this fact seems to me so important that, in a treatise such as the present, it is well, even in our terminology, to keep it constantly in view.

47. (*g*) This is, perhaps, the place to mention an objection to the conception of matter as consisting of force-centres, which has been brought by no less an authority than Prof. Clerk-Maxwell. He observes—"Theory of Heat," p. 86—that such a conception takes no account, and can take no account, of the property of inertia, which is essential to the idea of matter. To this the reply, on the definitions here laid down, is easy. No doubt the idea of inertia does not flow directly from the definition of force; but it does flow at once from that definition, combined with a general principle which nobody will dispute, namely that any effect which we can see and measure—and therefore any motion, amongst other effects—must be *finite*, and not infinite. Hence, since we define forces as the cause of motion, and since we know them only through the motions they cause, it follows that any forces we can investigate

must produce only finite motions in the bodies, or centres, to which they are applied; in other words, any known body under the action of any known force will only be caused to move over finite distances in finite times. And that is pre-cisely the fact which is expressed by the term inertia.

48. (*h*) It remains to answer the objection that I have no right to assume that the forces connected with matter are central forces at all. To this I reply that the conception of central forces is of course only one among many possible conceptions as to the action of forces, with which the student becomes perfectly familiar in his progress through mechanics. But the assumption that all the forces of nature are central forces is justified by the fact that all the natural phenomena, which have as yet been explained on mechanical principles, have been explained on that hypothesis. To cite only one instance, the widest generalisation yet made as to the mechanics of the universe, namely, that known as the Conservation of Energy, has been proved only on the sup-position that all the forces concerned are central forces. If anyone doubts this, he may be referred to the mathematical demonstration of the principle in any work on dynamics; or to explicit declarations such as that of Maxwell, "Theory of Heat," p. 93, or of Clausius, at the end of his demon-stration of the principle, "Mechanical Theory of Heat," p. 16. The latter is as follows:—"The assumption lying at the root of the foregoing analysis, viz., that central forces are the only ones acting, is of course only one among all the assumptions mathematically possible as to the forces; but it forms a case of peculiar importance, inasmuch as all the forces which occur in nature may apparently be classed as central forces."

49. (*i*) To some minds the definition here given of matter will be an offence, owing to an *à priori* impression that the action of two centres on one another across empty space cannot possibly exist. To this I can only reply as follows: —(1) *A priori* impressions have in all ages been the worst foes of science, and she has met and conquered too many to stop her course for any of them now. (2) This special conviction is probably not even a frequent, much less a general one; to many minds the idea of action across empty

space, or action at a distance, presents no greater difficulties than any other mode of action. (3) I have elsewhere shown —*Phil. Mag.*, January 1881—that various simple phenomena cannot be explained, except on the hypothesis that action at a distance exists ; and also that* the ordinary assertion of Newton's having disbelieved its existence is erroneous. (4) The definition does not absolutely preclude such minds from holding that the actions of matter are caused by the contact of some unknown description of extramundane particles, in any case where they can show that all the facts are reconcileable with that hypothesis. It seems clear, however, that for a very long time to come it will only be possible to represent mechanical facts to a student by telling him that the forces act *as if they were* central forces ; and while that is so the present definition will at least be the most convenient.

50. We have thus arrived, so far, at the following definitions :—(1) Mechanics is the science of motion, rest, and force. (2) Motion is an ultimate conception, which cannot be further defined, but is recognised by change of position with reference to something assumed to be fixed : all motion that can be recognised is therefore relative. (3) A force is a cause of motion. (4) Matter, or that which is moved by forcce, consists of a collection of centres of force distributed in space, and acting on each other according to laws which do not vary with time, but do vary with distance.

51. Having thus defined the science, and defined the special things with which it is concerned, the next step is to consider how these things are to be measured ; for if we hesitate fully to accept the well-known apophthegm, "Science is measurement," we cannot at least feel any doubt as to its converse, "No measurement, no science."

52. The mode in which motion, force, &c., are practically measured is known to all ; and there is no difficulty in showing how this mode is deduced from our definitions. We must begin by pointing out that there are three fundamental elements of all sciences, on the measurement of which measurements of every other kind are eventually founded.

* This, I find, has already been pointed out by Sir E. Beckett, "Origin of the Laws of Nature."

These are time, space, and weight. For the purposes of this inquiry I shall assume that these, as things, are perfectly well known, and shall not attempt to define them, if even such definition were possible. I need only in passing state that, like all ordinary men, I look upon them as things, and not, like a few bewildered metaphysicians, as abstractions, sequences, or what not. Passing on to consider how these are measured, we find in practice that the only possible mode is to adopt some one definite standard, kept at a definite place, which the inhabitants of a State agree to regard as a final appeal and measure on each of these subjects. Thus, in England the standard of time is the rate of a particular clock, the standard of space is the distance between two marks on a particular bar of metal, and the standard of weight is the weight of a particular mass of metal—all of which may be seen and used at Greenwich Observatory. It is, of course, essential that these standards should themselves be permanent and unalterable, and to ensure this they are frequently checked by various natural standards; e.g., in the case of time, for which the artificial standard is most liable to variation, it is checked by daily transit observations of sidereal time. Into these questions, however, we need not enter. Assuming that, by means of these artificial standards, we have the power of measuring with considerable exactness space, time, and weight, we have to inquire how we can apply these measurements to the purposes of mechanics; in other words, to the measurement of motion, force, and matter.

53. *Measurement of Motion.*—I have already observed that we cannot speak properly of the quantity of a motion, but only of its intensity, and that to the intensity of motion has been given the special name of velocity. Now, when we say that one body has a greater velocity than another, we simply mean that it is passing through a greater distance than the other in the same time. But space and time are both capable of measurement. Hence, if we can measure any interval of time, during which we know that the velocity remains constant, and can also measure the interval of space which a body travels over during that time,* then the ratio

* If we measure the space by the distance between two points, as is usually the case, we must take care to see that these points have neither

of the space to the time, as thus given, is a proper measure of the velocity of the body. If we are sure that the velocity does not alter appreciably during a unit of time, *e.g.*, one second, it is convenient to take that as the interval of time in all cases; and we then measure velocity by the space passed over in one second. But in practice nearly all velocities are constantly varying, and then we must adopt the usual mathematical expedient, and assume an interval of time so short, that the variation of velocity in that interval vanishes. If s be the symbol for space, and t for time, the symbolical expression for velocity will then be, by the ordinary notation of the differential calculus, $\frac{ds}{dt}$.

54. *Measurement of Matter.*—With a view to this object, Newton, in the introduction to the "Principia," introduced the well-known conception of "mass," and in this he has been followed by all subsequent writers. Probably many students have felt some difficulty in studying this conception, from the apparent obscurity as to its measurement. Mass, according to Newton, is the quantity of matter in a body, and is the product of its volume and its density. As volume and density are both things of which we have a tolerably clear conception, this language assists us, no doubt, in arriving at an idea of mass as a *thing;* but as we have no direct means of measuring density, it fails to establish a definition of mass as a term. Newton accordingly states explicitly that mass is to be measured by weight; but he gives nothing but experimental evidence to support this statement. On our definition of matter the question becomes perfectly clear. Matter being defined (for the purposes of this treatise) as a collection of like centres of force, the quantity of matter in a given volume simply means the number of centres of force contained in it, just as the quantity of shot in a given bag simply means the number of pellets. This then would be the *absolute definition* of mass; and the absolute mode of measuring mass would be to

of them altered their position, during or since the motion, with reference to whatever is assumed as absolutely fixed, *e g.*, in terrestrial measurements, the earth on which we stand.

count the number of centres. But from their immense multitude and smallness this is impossible, and some more practical method is to be sought for. Now since all centres act on each other alike, and with forces which are the same at the same distance, it follows that if we can isolate a point, or centre of force, and place it so that it is at practically equally distances from all centres in a body, then the force which the body exercises on this point will be simply the force due to one centre multiplied by the number of centres, and therefore will be proportional to that number. The force exercised between the point and the body would then be a measure of the body's mass. Now we cannot isolate a single centre, but practically we can treat the earth as a standard isolated body for this purpose, and we measure the mass of any other body by the force existing, under standard circumstances, between it and the earth—in one word, by its weight.

55. *Measurement of Force.*—We have seen that the solar system, considered mechanically, is a collection of like centres of force or points, acting on each other by like laws, which vary with the distance. Let us now consider a single point, as related to any number n of other points, which are all at the same distance from it, and form, in fact, an element of the surface of a sphere, of which it is the centre. It is evident that the force existing between any one of these points and the first point will be the same ; and therefore the total force acting between the first point and the element will be proportional to the number n of points, *i.e.*, by the last definition, to the mass of the element. If these forces are left to themselves, then, taking the central point as fixed, the element (since force causes motion) will move towards it or from it with ever-increasing velocity. But suppose another force to act directly upon the element and keep it at rest, or moving uniformly; then this force must balance the force existing between the element and the centre, and therefore it must be proportional to the mass of the element. Hence we see that where the velocity is unaltered force varies as the mass. But we have already seen, from the definition of force, that when the mass is unaltered force varies as the velocity caused in a given time.

Hence, by the ordinary law of proportion, when both are altered, force varies as the mass acted upon multiplied by the velocity caused in a given time. Now the product of the mass by the velocity of a body is called the Momentum, and forces when they act to produce motion in bodies are called Moving Forces; and we may therefore state the conclusion at which we have arrived by saying that moving forces are measured by the momenta which they generate in a given time. This is the principle laid down by Newton in Definition VIII., and in many later works announced as the Third Law of Motion.

56. On this system, the proper, or absolute, mode of measuring force would be to isolate two centres of force, place them at the unit of distance from each other, and observe the value of $\frac{ds}{dt}$ at the first instant after they began to move. This would form an absolute elementary unit of force; and the force acting between a centre of force and a body, in any other case, would be found by multiplying this absolute unit by the number of centres in the body—in other words, by its absolute mass—and by the function of the distance which expresses the law of the forces. It is obvious that this process is impossible. As in the case of mass, we must resort to the standard furnished us by the earth on which we live. We begin by taking a certain piece of matter—such as the standard pound at Greenwich—which we agree to regard as a unit of mass. The weight of this pound, taken at Greenwich—since the force exercised by the earth may differ at different places,—we take as the unit of weight. Any force, so long as it produces no motion, is measured by the number of such pounds which it will support—in other words, by the number of units of weight which it will balance. This is the measure of Statical Force, and its unit is the weight of a pound.

57. Again, if we consider forces as acting always on a unit of mass, and if we suppose that there is no force acting in the opposite direction, then these forces will be measured simply by the velocities which they generate in a given time. This is the measure of Accelerating Force, and its unit is a unit of velocity generated in a unit of time. In England

the unit of velocity is 1ft. per second, and since gravity, or the attraction of the earth, generates a velocity of 32·2ft. per second in one second, we say that the value of the accelerating force of gravity is 32·2.

58. Thirdly, if forces act on different masses, and produce motion in them, then the forces are measured by the product of the mass and the velocity, or by the momentum generated in a unit of time. This is the measure of Moving Force, and its unit is a velocity of 1ft. per second, generated in a mass of 1lb. weight. Here, as before, we must consider that there are no forces acting in the opposite direction. In other words, the measure of moving force is only the measure of the unbalanced part of a force. The balanced part of a force is to be measured by statical standards only.

59. It remains to say a word about the relations of Mass to moving force. Since the moving force of gravity is measured in any particular case by the mass moved, and by the velocity generated in a second—which latter is 32·2, or g—the moving force of gravity on a mass M will be represented by $M g$. But the statical and moving force must clearly be proportional to each other, since they are the same things acting in different ways; and hence, if W be the weight of the body in pounds, we must have $W = C\,Mg$, where C is a constant. It is convenient to make this constant unity, so that we may write $W = M g$; and this is done by assuming the unit of mass—which has not been fixed—to be the mass of 32·2 lb., instead of simply the mass of 1 lb., as would otherwise seem more natural.

60. *Laws of Motion.*—We have now defined the things with which our science is mainly concerned—Motion, Force, and Matter—and have shown how each of these is in practice measured with sufficient accuracy to make a true science of them possible. We are thus at last in a position to go further, and consider the leading principles of that science itself.

61. Newton, when he arrives at the same point in the "Principia," proceeds by laying down three "Axioms or Laws of Motion," which, though not actually discovered by him, have ever since borne his name. These, literally translated from his own words, are as follows :—

" 1st *Law of Motion.*—Every body continues in its condition of rest, or of uniform motion in a straight line, except in so far as it is compelled by impressed forces to change its condition.

" 2nd *Law of Motion.*—Change of motion is proportional to the moving impressed force, and takes place along the straight line in which the force is impressed.

" 3rd *Law of Motion.*—Reaction is always opposite and equal to action ; or the mutual actions of two bodies are always equal, and in the opposite directions."

62. Newton gives these laws as independent axioms, and merely adds proofs, or rather illustrations, drawn from experience. But in the case of the first I have already shown that it flows directly from the definitions of force and motion, combined with the universal truth which has been called the Principle of Conservation. This places the law on a much more satisfactory basis than it occupies when regarded as an independent fact, proved only by experience. I have now to show that the same or a similar basis may be found for the other two laws. As enunciated, the second law of motion seems at first sight to be nothing more than the principle that force is proportional to the velocity which it generates, expressed in another form. But Newton adds to the enunciation a very important scholium or explanation, which may be translated as follows :—" If a force generates a certain motion, then double the force will generate double that motion, triple the force triple that motion, and that whether it be impressed at the same time and instantaneously, or gradually and successively. And this motion (since it is always directed in the same straight line as the generating force), if the body was already in motion, is added to the previous motion where it is in the same direction, or subtracted from it where it is in the opposite direction, or is obliquely combined with it if the directions are oblique, and is compounded with it according to the direction of each."

63. It will be seen that Newton here contemplates, as a fact involved in his second law, though not explicitly stated, that each force acting on a body will produce its full proportionate effect of motion, whatever other forces or motions the body may be subjected to at the same time ; and this

fact has naturally come to be looked upon as the new and essential truth contained in the second law, so that it has been gradually included in, and finally has entirely usurped, the enunciation of the law, which, in Goodwin's " Course of Mathematics," for instance, stands as follows :—" When any number of forces act upon a body in motion, each produces its whole effect in altering the magnitude and direction of the body's velocity, as if it acted singly on the body at rest."

64. At first sight this may appear to be something like a truism. If forces are only known to us by their effects, *i.e.*, the motions they cause, then a force, or any part of a force, which in consequence of the motion or other conditions of a body does not produce its effect, is the same to us as if it did not exist—in other words, if a force is recognised at all, it must produce its effect, because it is only by the effect that it is recognised. But this view is doubly fallacious. In the first place, there are forces which we recognise otherwise than by their effects, namely the forces which we ourselves exert. And in the second place, the law practically asserts that there are constant forces, always acting alike, and whose effects are the same whether the matter on which they act be in motion or at rest, and whether or no other forces are also acting upon it. For instance, it asserts that the weight of any body is such a constant force, and will always have its full effect ; so that a bullet will fall exactly as far towards the earth in one second, whether it be simply dropped from the hand or fired horizontally out of a rifle. This is, no doubt, an essential point to be cleared up, before any problems which involve the action of force are considered ; but when thus stated, it is at once seen that it is a necessary consequence of our definitions, and of the principle of conservation. For if the law did not hold—if a force were hindered from having its full effect by circumstances, such as the presence of other causes—this could only take place in two ways : either the force would not act so as to produce its effect, or the effect, being produced, would of itself die away. Now the first of these suppositions is against our definition of matter, which asserts that the forces do not vary with the time, and must, therefore, be always the same if the distances be the same ; and the second is against the principle of con-

servation, which asserts that effects live. Hence the second law of motion is not an independent principle, but a necessary consequence of facts already arrived at.

65. The same may be said of the Third Law. Since any centre of force acts on any other with a force which is always the same if the distance be the same, and since the distance between two centres must be the same in whichever direction it is measured, it must follow that if either of these be called the Action, then the other, or the Reaction, must be equal and opposite to it. The same holds true of the action between any centre and a group of centres, or a body; and similarly of the actions between one group of centres, or body, and another. Thus the Third Law follows at once from the definition of matter. But it is necessary to point out, even at the outset, that we must be capable of analysing the action between any two bodies into the actions between the individual centres of which they are composed, and of calculating the sum of these actions, before we can say what the net action and reaction between two given finite bodies really is; and this requirement, which can rarely be fulfilled in practice, makes much caution requisite in the application of the law. Thus when a shot strikes a target, the forces of action and reaction induced by the shock are no doubt equal and opposite; but their exact nature and distribution who shall calculate? Most of the cases which are cited as illustrations of the law are so under special circumstances only. Thus Newton himself uses the illustration of a horse drawing a cart; but in that case the reaction and action, as represented by the pull at the two ends of the trace, are equal only when the horse and cart are alike moving with uniform speed. If, for instance, it were true at the commencement of the horse's effort, so that his pull at one end of the trace was always counterbalanced exactly by a resistance at the other, it is clear that the start would never take place at all. As a matter of fact, the traces are alternately tightening and slackening continually, and the times during which the speeds of the horse and cart are really the same are probably indefinitely short. The other illustration used by Newton, which is that of the finger pressing a stone, is less liable to mistake, since then no

motion takes place; and, in fact, it is only in the domain of statics that the law can be used without some circumspection.

66. *Accelerating Forces.*—The laws of motion, with our definitions, enable us to lay down at once the fundamental propositions as to the action between two centres of force, or between two bodies each of which may be supposed to be concentrated in a single centre. Practical examples of such cases are the attractions of the sun and earth, neglecting the disturbing forces of the planets, or the fall of a body to the earth, neglecting the resistance of the air. In such cases the ideas involved become greatly simplified. In the first place, since all motion is relative to some point assumed to be fixed, we shall naturally assume as our fixed point one of the two centres of force concerned, and thus investigate the motion of the other in reference to it. Thus, in the case of the earth and sun, we assume the centre of the sun as fixed; and in the case of a falling body, we assume the centre of the earth as fixed. Secondly, if the body assumed to be in motion be a single centre of force, it is an absolute unit of mass; and if it be a group of centres, yet they will all move as one, and therefore the idea of mass need not be included, at least as long as the two bodies are still at a considerable distance from each other. Thirdly, as no other forces are acting, the whole of the action will take place in the straight line joining the centres, and the question of the combination of forces, which has not yet been settled, does not enter.

67. Let us consider two bodies, A and B, and investigate the motion of A with regard to B, taken as fixed. The problem will be stated mathematically thus:—*Given the distance* l *between* A *and* B, *at the instant when* A *begins to move, and the accelerating force* f *with which* B *acts on* A; *to find the velocity of* A (*referred to* B) *at* t *seconds after that instant, and also the space* s *which it has passed over from its original position towards* B.

68. Let us, in the first place, assume that the force with which B acts on A is constant. This is, of course, against our definition of matter, and is never exactly true in nature. But it may be assumed as true—according to the ordinary

principles of mathematical reasoning—when either the interval of space or the interval of time is, relatively, exceedingly small. The former of these suppositions holds practically in the case of falling bodies, where the variation of gravity due to the approach towards the centre of the earth may always be neglected. The latter will enable us to calculate the effect of a varying force by the ordinary methods of the integral calculus.

69. Let us then begin by assuming that the force is constant. It will therefore be measured by the velocity f generated in one second (Art. 57). Moreover, by the second law of motion, this velocity will in no way affect the action of the force, which in the next second will generate exactly the same velocity f. At the same time, by the first law, the velocity f, generated in the first second, will remain unaltered. The velocity at the end of the second second will therefore be $f+f$, or $2f$. In the third second a third velocity f will have been added, making the velocity $3f$. By similar reasoning the velocity at the end of t seconds will be tf; or, if V be the velocity of A at the end of t seconds, we have

$$V = ft \quad . \quad . \quad . \quad . \quad . \quad . \quad (1).$$

Next, to find the space s described in the time t. With the aid of the integral calculus, this is easily accomplished, as follows:—We have already seen (Art. 53) that velocity is measured by the limiting value of the ratio of space to time; or, if v be the velocity at any instant—

$$v = \frac{ds}{dt}.$$

But from above $v = ft$. Hence—

$$s = \int v\,dt = \int ft\,dt = \frac{f\,t^2}{2} \quad . \quad . \quad (2).$$

No constant is added, because $s = 0$ at the beginning of the motion, or when $t = 0$. Combining the two, we have,

$$V^2 = f^2\,t^2 = 2fs \quad . \quad . \quad . \quad . \quad (3).$$

70. To extend this to the case of varying forces, such as alone exist in nature, we may assume the above equations

to hold for indefinitely small intervals of time, during which all the conditions are constant. Hence we may put

$$d v = f d t \quad . \quad . \quad . \quad . \quad (4),$$

$$v = \frac{d s}{d t} \quad . \quad . \quad . \quad . \quad . \quad (5);$$

whence

$$v = \int_0^t f \, dt;$$

or

$$\frac{d s}{d t} = \int_0^t f \, dt;$$

whence, differentiating,

$$\frac{d^2 s}{d t^2} = f \quad . \quad . \quad . \quad . \quad (6).$$

These are the ordinary equations of analytical dynamics, which must then be dealt with as described in any of the text-books on that subject.

71. The following proof of equation (2) above may be given for the benefit of those who are not acquainted with the integral calculus. Suppose the time t to be divided into a very large number n of very small equal intervals w. Then we may suppose that during each of these intervals the velocity remains constant at the value it has at the beginning of the interval; and the smaller the intervals, and therefore the larger their number, the nearer will this supposition be to the actual truth. But if the velocity be constant, the space described in each interval will be simply the velocity × the interval. Now, by equation (1), the velocities at the successive intervals, 1 to n, are $0, f \times w$, $f \times 2 w, f \times 3 w, \ldots f \times (n-1) w$. Hence the spaces described in the successive intervals are $0, f \times w^2, 2 f \times w^2$, $3 f \times w^2, \ldots (n-1) f \times w^2$. But the total space must be the sum of the spaces described in the intervals, and it is therefore equal to $f w^2 [1 + 2 + \ldots + (n-1)]$, or to $f w^2 \times \frac{(n-1) n}{2}$. But $n w = t$; hence this may be written

$$\frac{f t^2}{2} - \frac{f t^2}{2 n}$$

The larger n is, the nearer will this expression be to the truth; but the larger n is, the smaller does the second term of the expression become. Hence in the limit this second term will vanish, and the true value of the space described in the time t is $\dfrac{f t^2}{2}$

72. *Composition of Forces and Motions.*—We have next to consider the effect of two or more forces, acting simultaneously on the same centre. This cannot be done by the aid of the three laws of motion alone, since the second law, or rather Newton's explanation of it, only decides the question where the two forces are in the same straight line. In that case the velocity of the centre, at the time t, will be the algebraical sum of the velocities generated by each force in that time; it being obvious that velocities, being measured in units of length, may be taken as positive or negative, added or subtracted, exactly as lengths are treated in analytical geometry. But suppose two forces to act on the same centre obliquely to each other; then the second law states that the effects are to be compounded together, but in what manner does not appear. To solve this question some additional p inciple is necessary. The principle generally used for this purpose is that a force may be supposed to act at any point in its line of action, provided that point may be treated as rigidly connected with the body in question. This principle, though of course true, is by no means self-evident; it involves conceptions, such as rigid connection, which are quite unfamiliar to a student when commencing mechanics; it introduces the vague and clumsy conception of a body, in place of the simple and accurate conception of a centre of force; and, lastly, it is a principle which is scarcely used or heard of on any other occasion, and is no whit wider than the proposition it is adduced to prove. For these reasons another and more satisfactory principle seems desirable, and such a principle, open to none of the objections alleged, may be found in the principle of Symmetry.

73. This principle, in its general form, may perhaps be stated thus :—When a cause, or set of causes, is so related to two opposite effects that there is no reason whatever why one of those effects should take place rather than the other, then neither of the effects will be produced by the cause,

or causes; and this relation is said to be a relation of symmetry.

74. Of this principle, as of others, it may be true that when thus stated in its general form it is not easy at once to grasp its bearings. An illustration or two will make it clearer. We put a rein or a curb on each side of a horse's mouth, and then by pulling on both together we know that we shall not cause him to turn to either side, because there is as much reason to turn to the one as to the other. In Euclid, the unexpressed axiom that figures which coincide are equal to each other, really rests on this principle, since there is no reason why one of two such figures should measure more than the other, or why it should measure less. In ordinary mechanical practice we admit at once that two equal weights suspended over a pulley by a weightless string will remain at rest, because there is no reason why either should either rise or fall; and for the same reason that two equal weights suspended from the ends of an equal-armed horizontal lever will also remain at rest. This last fact has indeed been made the basis of a complete system of statics —see Goodwin's "Course of Mathematics," p. 225—which would thus rest directly on the principle of symmetry. Looked at in the light of these illustrations, it may perhaps be thought that the principle should be regarded as a corollary from the great principle of causation; the cause, being as likely to produce one effect as the opposite, is really not a cause tending to produce either; and without a cause there will be no effect. I do not care to question this way of looking at the principle; but I believe it is at least equally good philosophy to regard the principle of causation as itself a deduction from a great number of observed facts, of which those of symmetry are some of the most important.

75. Original or derived, the principle of symmetry is one which nobody has ever cared to deny, and we may, therefore, accept it as an unquestioned truth, and apply it at once to the simplest case of compounding forces, which is the case of forces at right angles.

76. *Problem:—A point or centre is acted on by two forces whose directions are at right angles; to find the resulting motion of the centre.*

77. Suppose the two forces to begin to act on the centre at rest, and to act upon it for an indefinitely small time dt, and then to cease. If we assume for the moment that only one force has acted, it will have generated in that time a certain velocity $f\,d\,t$, where f is the measure of the force, being the velocity which would be generated if the force continued constant for one second. Since the time considered is indefinitely short, we may consider the force as constant during that time. This velocity $f\,dt$ will cause the centre to describe, in the next element of time, a certain indefinitely small space ds, proportional to the velocity $f\,dt$. Hence ds is proportional to f. Let P be the position of the centre at the end of the first interval $d\,t$, and let P Q be the element of space which would be described in the second interval. Then P Q is proportional to f. Let P R be the element of space which would be described in the second interval, supposing the other force f^1 to be the only one acting. Then by similar reasoning P R is proportional to f^1, and by hypothesis P R is at right angles to P Q. We have now to find the space actually described by P under the joint action of the two forces.

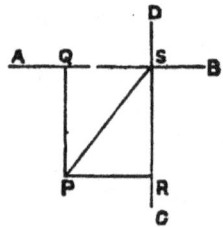

78. By the second law—Art. 63—f will still produce its full effect, and will therefore actually cause P to describe the space P Q, except in so far as that space is increased or diminished—in other words, in so far as P is accelerated or retarded—by the action of f^1. But since the direction of f is at right angles to P Q, it has no more tendency to produce an acceleration of P along P Q than to produce a retardation, and *vice versâ*—in other words, f^1 is symmetrical with respect to motion along P Q, and, therefore, by the principle of symmetry, it will produce no effect either in accelerating or retarding. Therefore P will still travel the full distance P Q in that direction. But by exactly similar reasoning, it will also travel the full distance P R in that direction; for by the same principle of symmetry, the fact that it is also moving in the direction P Q will have no effect on its motion in the direction P R.

This amounts to saying that, at the end of the time dt, P will be found somewhere on the line A Q B, which is drawn through Q at right angles to P Q; and also somewhere on the line C R D, which is drawn through R at right angles to P R. Hence its actual position must be the point S in which these two lines meet. And since the elements of space here considered are indefinitely small, the path of the centre will not differ from a straight line joining its extreme positions; in other words, from the straight line P S. But P S is the diagonal of a rectangle, the sides of which are proportional to f and f^1. Hence, P S is similarly proportional to $\sqrt{f^2 + f^{12}}$; and, therefore, the path of the particle is exactly the same as if it had been acted on during the first interval $d\,t$ by a force $\sqrt{f^2 + f^{12}}$, whose direction coincided with P S. Hence, we have this result:— *When a centre is acted upon by two forces at right angles to each other, the effect is exactly the same as if it were acted upon by a single force, which is represented in magnitude and direction by the diagonal of any rectangle, whose sides represent in magnitude and direction the two forces acting.*

79. In the foregoing proof we have supposed for simplicity that the forces cease to act during the second interval ; but by the second law the action of the forces during that interval will not alter the effect due to the action of the forces during the first interval, but must simply be added to or subtracted from it. But, by similar reasoning, this action in the second interval will be the same as if the diagonal force $\sqrt{f^2 + f^{12}}$ continued to act during that interval; and therefore the result which we have arrived at will hold also for the second interval, and also for the third, fourth, fifth, &c., and is generally true for any time during which the two forces continue to act on the centre.

80. The above proof gives the principle of the combining, or, as it is usually called, the compounding of two forces, at right angles to each other, which act together on any centre. Conversely, if a centre is acted on by a single force, P S, we may in thought regard it as under the action of two forces in any two directions, P Q, P R, at right angles to

each other, provided we consider the values of these forces
to be represented by P Q, P R, the sides of the rectangle of
which P S is the diagonal. On this supposition we are said
to resolve the single force into two component forces at right
angles to each other.

81. We have now to extend the proposition to cases where
the forces are not at right angles.

82. *Parallelogram of Forces.—Problem :—A centre is acted
on by two forces whose directions are at any angle with each
other ; to find the resulting motion of the centre.*

83. Precisely as in the last proposition, we may represent
the effect of the two forces f and f^1, considered singly, by
two straight lines P Q, P R, drawn
from the position of the centre
at the end of the first interval
dt. Complete the parallelogram
P Q S R, and join P S. Draw Q A,
R B, perpendicular to P S, and complete the parallelograms C Q A P,
D R B P. Then, by article 80, we
may consider the force P Q as
resolved into the two forces P C, P A,
at right angles to each other, and the force P R as resolved
into the two forces P B, P D, at right angles to each
other. But by geometry it is evident that A Q = B R, and
therefore P C = P D; hence the force represented by P C is
equal in magnitude and opposite in direction to the force
represented by P D; and therefore by the principle of symmetry they will cancel each other, and will have no effect
upon the motion of P. Hence the motion of P will be
entirely due to the forces represented by P B and P A, and
therefore it will be along the common direction of those
forces, that is along the line P S. Moreover, by geometry,
P B = A S ; therefore, P A = B A + A S = B S, and therefore
P B + P A = P S. Hence the joint effect of the forces f and
f^1 will be to make the centre P move through the space P S,
which is the same as the effect of a single force represented
by P S. Hence, just as before, we establish the general
proposition:—*If a centre is acted upon by two forces at once,
whose directions make any angle with each other, the effect*

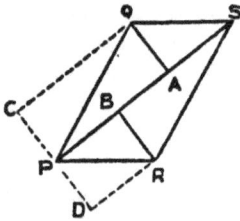

*will be the same as if it was acted on by a single force, which
is represented in magnitude and direction by the diagonal
of a parallelogram, whose sides represent in magnitude and
direction the two forces acting.*

84. We have thus established the proposition of the
Parallelogram of Forces, which forms the foundation of
statics; and that branch of mechanics can thenceforward be
studied in the ordinary manner, without the introduction of
any new principles. It should be observed that the forces,
in the proof of the Parallelogram of Forces, have been represented by the velocities generated by them, as in dynamics,
and not by the pounds weight they will balance, as in statics;
but it is clear that a proposition which is true of things when
measured in one way cannot become false when they are
measured in another way, and therefore the proposition may
be at once assumed to be true for the science of statics, as
well as for that of dynamics.

85. *Parallelogram of Velocities.*—From the fact mentioned
in the last section, that in the proof of the Parallelogram of
Forces the forces have been represented by the velocities
generated by them, it follows at once that the velocity of a
centre at any moment may be resolved, in thought, into two
component velocities, on exactly the same conditions as a
force may be resolved—viz., that the lines representing the
components form the two sides of a parallelogram, of which
the diagonal represents the actual velocity. This proposition
is usually demonstrated independently as the Parallelogram
of Velocities, and is often stated in the converse form,
namely, that two velocities, existing at the same time, may
be compounded into a resultant velocity. This, however, is
a confusion of ideas. It is quite right to regard a body as
under the action of two forces at the same time, and to
compound their effects, because such forces are separate
realities; but the velocity of a body at any instant must be
in one definite direction, and of one definite amount, and it
is only in thought that it is possible to analyse it into two
velocities tending in two different directions, and capable of
being studied independently.

86. *Moving Forces—Energy—Work.*—We have now to
consider the principles to be followed in dealing with moving

forces. . The fundamental question in this department may be stated as follows:—*How are we to measure the total effect of a given force, when it has acted for a given time on a centre, or a group of centres of given mass, which is in motion during the action?*

87. The answer to this question may be stated at once. It is that the effect is measured, with regard to the force—which for the present we may consider to be constant—by the product of the force and of the distance moved through by the centre or body, in the direction of the force, during the action. And this product is called the Energy exerted by the force. Again, with regard to the body moved, the effect is measured, if the body start from rest, by the product, at the end of the action, of the mass and of half the square of the velocity in the direction of the force; or by the difference between this product and the similar product taken at the beginning of the action, if the body has already a velocity in that direction when the action commences. This product is called the *vis viva*, or the kinetic energy, or the actual energy, or the energy of motion, of the body; and the effect of the force is therefore measured by the change in this quantity, whichever name be used for it, during the action. Throughout this paragraph it is assumed that there is no force acting on the body in the opposite direction.

88. These principles are laid down in all text-books, but usually without any explanation of the reason why the energy exerted—*i.e.*, the product of the force and the distance—is the proper measure of the effect of the force. Nor does it at first sight seem clear why the element of time should be altogether absent from the measure of the effect, or why this measure, referred to the body, should be in terms of the square of the velocity, and not of the velocity simply.

89. To elucidate this, let us suppose that the force, instead of being continuous throughout the motion, acts discontinuously at certain equal small intervals of space *ds*, so that, at the beginning of each of these intervals, there is an instantaneous action which generates in the body exactly the same velocity as is really generated during that interval by

the continuous force. Let the number of these intervals in any space s be n, so that $s = n\,ds$. Then the total effect on the body while traversing the space s will, by the second law of motion, be the sum of the n effects due to the action of the n equal impulses at the beginning of the n intervals. It is evident, therefore, that, so long as the strength of the impulses remains the same, the total effect will vary as n; or, since $s = n\,ds$, and ds is supposed always the same, the total effect will vary as s. Again, if the number of the impulses remains the same, the effect will of course vary with the strength of each impulse ; in other words, with the force. Hence, by the ordinary principle of variation, if both the spaces and the forces be different, the effect will vary as their product. But by considering the length of each interval ds as indefinitely small, and therefore the number n as indefinitely great, we may make the assumed circumstances approach indefinitely near to those of a constant force acting continuously over the same space ; and hence we may say that the effect of such a force will vary as the product of the force itself, however measured, and the distance through which it acts. This, as already stated, is called the energy exerted by the force. It does not contain the element of time, because neither the number of impulses nor the strength of each impulse are in any way affected by the velocity with which the body passes over the successive intervals ds ; in other words, by the time which it occupies in describing the whole distance s.

90. We have thus shown that if F be the moving force of a given centre A, and s the space through which it acts on any other body or centre B, its effect will be properly measured by the product F s. This assumes that F is constant. If F vary with the distance, as will always be true in nature, then the same holds for each indefinitely small element of space ds—i.e., the effect of the force, while the moving centre B traverses this element, is measured by Fds. To find the effect for any finite space s, we have only to integrate F ds from O to s.

91. Let us now express the same product in terms of mass and velocity. Since the force is a moving force, it is properly expressed (Art. 58) by M f (where f is

the velocity generated in one second), or (Art. 70) by
$M\dfrac{d^2s}{dt^2}$. Also, since ds is indefinitely small, we may
consider that the velocity of B, while it traverses ds, is
constant; hence we may write (Art. 70),

$$ds = v\,dt = \frac{ds}{dt}\,dt.$$

Hence, for F ds we may write—

$$M\frac{d^2s}{dt^2}\frac{ds}{dt}\,dt;\ \text{or}\ M\,v\frac{dv}{dt}\,dt.$$

This we have to integrate between the limits o and t, where
t is the value of the time when the space s has been
described. The value of this integral is well known to be
$\frac{1}{2}$ M $(v^2 - v_0^2)$, where v_0 is the value of the velocity when
$t = 0$. Hence it appears that the effect of the force may
also be represented by $\frac{1}{2}$ M $(v^2 - v_0^2)$—that is, by the
change in the kinetic energy, *vis viva*, or whatever other
name we apply to that quantity.

92. The two modes of representing the effect of a continuous force—by energy exerted and by the change in the
vis viva—are thus established. The resolution of the single
resultant force, so as to extend the principle to three dimen-
sions, and thus make it general, may easily be accomplished,
as shown in any of the ordinary text-books.

93. Hitherto I have assumed that there is no force acting
on the moving centre B in the opposite direction to its
motion towards the fixed centre A. This is what would be
true if A and B were the only centres in the universe. In
nature such a case, of course, cannot occur. The number
of centres in the universe is incalculable, and, by our definition of matter, these all act upon both A and B. There are
cases, however, where the action of these extraneous centres,
owing to their distance or other causes, is insignificant when
compared with the direct action between A and B, and may
for many purposes be neglected. One such case is that of
a body falling to the earth in vacuo. Another is that of a
planet revolving round the sun, where the small disturbances
due to other planets may, for many purposes, be neglected.
In these cases the effects are represented with sufficient

exactness by the expressions indicated above. But I must now go on to consider what modifications are introduced by the presence of other forces.

94. Let me as usual take the simplest case, and assume that, in addition to the fixed centre A and the moving centre B, there is a third centre C placed in the prolongation of the line A B, on the other side of B, and therefore acting upon B in the opposite direction to the action of A.* For further simplicity we shall assume (1) that C and A have no mutual action; (2) that B is initially at rest; (3) that C as well as A is fixed; (4) that the forces with which B and C act on A are constant, or only vary by amounts that may be neglected. We will consider hereafter how far these assumptions can be realised in the universe, and how far they affect the conclusions.

95. Let P and Q be the forces, measured dynamically, with which A and C respectively act upon B; and let P be greater than Q. Then, by the second law of motion (Art. 63), each of these forces will produce its full effect exactly as if the other was not present, and the net effect upon B will be simply the difference between these opposite effects. Let s be the distance through which B has moved towards A at the end of a given time. Then the effect of A will be measured just as before by the product P s. For as in our former proof (Art. 89), we may imagine the action of A divided into impulses, which act at successive points in space separated by small intervals, but which are independent of the time occupied in describing the intervals between one point and another—as they must be, because, by our definition of matter, the forces are not functions of time. Each of these impulses will produce its full effect, by the second law, independently of the action of C; and, as before, the total effect will vary jointly as the number and strength of the impulses, and will, therefore, be represented by P s. Let us now make a similar assumption with regard to the action of C, namely, that it is divided into impulses acting at the same points in space as those of A. Each of

* It is supposed throughout that the forces are attractive; if they are repulsive the demonstrations will not be affected, but B's motion will be in the opposite direction.

these impulses will produce its effect, but this effect will be neutralised, as regards B, by the opposite impulse due to A; and the net impulse actually given to B will be the difference between the impulses due to A and C respectively. Since these impulses all vary as the forces, *i.e.*, as P and Q respectively, it follows that the net effect on B will be exactly the same as if it had been acted on by a single force $P - Q$. Taking this for the value of the force, our former investigation (Art. 89), will hold again; the energy exerted on P will be represented by $(P - Q)s$, as regards the force, and as regards the body B it will be represented by the *vis viva*, or $B v^2$, where B is taken for half of B's mass, and v its velocity after describing the space s.

96. The facts regarding the real effect on a body of two opposing forces have now been set forth ; it still remains to consider the names by which the different quantities involved are to be designated. It will be found that names are required for the following :—(1) The force P with which A acts on B, and which actually causes the motion of B in the direction B A ; (2) the force Q with which C acts on B, which tends to cause motion of B in the direction B C, and actually retards its motion in the direction B A ; (3) the difference between these forces, or $P - Q$, which is the net force acting upon B, and causing its motion ; (4) the total effect of A, which, as we have seen, is represented by Ps ; (5) the total effect of C, which is represented by $-Qs$; (6) the net effect of the action of A and C together upon B, which is represented either by $(P - Q)s$, or by $B v^2$. To some of these we have already assigned names provisionally, but it will be well to go through them all.

97. With regard to the first three, I shall have no difficulty in adopting the nomenclature of Rankine, which appears to be the only one definitely proposed. We shall thus give to P the name of the effort, to Q that of the resistance, and to $P - Q$ that of the unbalanced effort. The first name is not quite free from objection, because in common parlance we speak of effort rather when we do not succeed in overcoming a resistance than when we do ; but it may be allowed to pass in the absence of any competitor. Again, No. 4, or the total effect of A, is everywhere known as the

energy exerted by A. So far all is simple. When we come to No. 5, or the total effect of C, the case is different. We are here face to face with a variation in nomenclature on the part of our highest authorities, which has not been generally noticed, and which is at least liable to lead to confusion.* On the one hand, Rankine gives to this effect, or $-Qs$, the name Work done. According to him, work is done only when resistance is overcome, and when, therefore, there is a third body acting, such as C, which tends to move B in the opposite direction to its actual motion. On the other hand, Thomson and Tait, Clausius, and most of our more recent writers on dynamics, give to the expression work done a much wider signification. With them it is in fact equivalent to the energy exerted, or Ps; being the same action, but looked at from the point of view of the body acted upon, rather than the body acting. The name they apply to $-Qs$ is not always well determined; but we shall assume it to be the Potential Energy imparted to B. The reason of this name we shall see hereafter. To No. 6, represented by $(P-Q)s$, or Bv^2, they would assign the name of the Kinetic Energy imparted, while Rankine would call it the Actual Energy; both these names being intended to supersede the older term of *vis viva*, which we have hitherto provisionally used.

98. It is necessary to prove the fact of this variation in nomenclature, by quoting and illustrating the exact words employed by the authors referred to, and I shall then comment briefly upon it.

99. In order to set forth Rankine's views, it will be best to give the following extract from his "Applied Mechanics," 1st edition, Art. 513:—"Work consists in moving against resistance. The work is said to be *performed*, and the resistance *overcome*. Work is measured by the product of the resistance into the distance through which its point of application is moved. The unit of work commonly used in Britain is a resistance of one pound overcome through a distance of one foot, and is called a foot-pound. Energy

* I called attention to this variation in a paper read before the Physical Society in March, 1881.

means capacity for performing work. The energy of an effort, or potential energy, is measured by the product of the effort into the distance through which its point of appli-cation is capable of being moved. The unit of energy is the same with the unit of work. When the point of application of an effort has been moved through a given distance, energy is said to have been exerted to an amount expressed by the product of the effort into the distance through which its point of application has been moved."

100. It may be argued that in these definitions Rankine intended to include the resistance—allowing for the moment the use of the term in that connection—of inertia, which would exist even where there are only two bodies concerned. That this was not the case (apart from the absurdity of speaking of 1 lb. of inertia) is made abundantly clear as follows:—

101. (*a*) Rankine's definition of resistance (Art. 511) is as follows:—"A direct force is further distinguished according as it acts with or against the motion of the point . . . . by the name of effort, or of resistance, as the case may be." Now inertia certainly cannot be an effort, therefore it cannot be a resistance.

102. (*b*) If the inertia be added to the resistance, and the sum considered equal to the effort, then energy exerted and work done are always equal, being, in fact, opposite views of the same thing. Of this Rankine gives no hint.

103. (*c*) Work is said to be measured by the resistance. But we have no measure of inertia, excepting the distance through which it is overcome by a given effort; therefore it must be the effort and not the resistance by which the work must be measured, if inertia is included in the latter.

104. (*d*) In Art. 549 (p. 499) Rankine takes the case of "a moving body acted upon by an effort P and a resistance R, the effort being the greater, so that there is an un-balanced effort P — R;" and he lays down the equation resulting as follows:—

$$\frac{W}{g}\left(v_2 - v_1\right) = (P - R)\,\Delta\,t.$$

It is evident that R does not here indicate the inertia. A yet clearer case is Art. 689, p. 622, on fluctuations of speed, where P and R—effort and resistance—are represented by different lines, and the work performed is measured by the value of R.

105. It remains to state the opposite view. On turning to Thomson and Tait's "Natural Philosophy," Part I., 1873, Art. 204, p. 63, I find the following definition :—"A force is said to do work if its place of application has a positive component motion in its direction, and the work done by it is measured by the product of its amount into this component motion. . . . In lifting coals from a pit, the amount of work done is proportional to the weight of the coals lifted ; that is, to the force overcome in raising them, and also to the height through which they are raised. The unit for the measurement of work adopted in practice by British engineers is that required to overcome a force equal to the weight of a pound through the space of a foot, and is called a foot-pound."

106. It is obvious that the definition of work in the first paragraph is different from that given by Rankine. On the other hand, it agrees with that of Clausius—"Mechanical Theory of Heat," p. 1—which puts the point as clearly, perhaps, as is possible. Assuming the force to act on a single material point, he proceeds : "If this point . . . travels in the same straight line in which the force tends to move it, then the product of the force and the distance moved through is the mechanical work which the force performs during the motion."

107. The existence of these two modes of defining work done is thus, I believe, put beyond possibility of doubt. It is clearly desirable that one of these modes should be suppressed, and the other definitely adopted, and it remains to decide which should have the preference. Arguing the question *ab initio*, the following reasons, on behalf of retaining Rankine's nomenclature, appear to have much weight.

108 (*a*) It gives a separate short and definite name for all the quantities concerned, especially if *vis viva* be retained for No. 5 ; this can hardly be said of the other system.

109 (*b*). Whichever system may be preferred for dynamics, there can be no doubt that Rankine's is the most convenient for applied mechanics. In dynamics the resistances are

usually deducted from the efforts at the commencement, and need not be afterwards considered. But in applied mechanics it is absolutely necessary to take account of the resistances. Thus, to vary a little the illustration given in Thomson and Tait's definition, the problem of moving 10 tons of coals is sufficiently stated for the mathematician, as soon as he knows that the coals start from rest, and that an excess of effort over resistance = 1 ton is available for moving them. He can calculate their velocity at the end of a given time equally well, whether they are standing on a level tramway, on which the resistance to traction is 1 cwt., or hanging in a shaft with their full weight of 10 tons. But this makes all the difference to the engineer who has to fix the strength of the rope, and design the engine which shall do the work.

110 (c). Rankine's system was used by him throughout his manuals, which are the recognised text books on the various branches of scientific engineering, and are constantly consulted and appealed to accordingly.

111 (d). The other set of definitions makes it necessary to regard inertia as constituting a form of resistance. Now, in common parlance we do not say that a stone, for instance, offers in itself any resistance to falling towards the earth; we reserve the term for the forces opposing its motion, such as the resistance of the air. And it seems also more philosophical to make a distinction in phraseology between the case where the motion is influenced by an actual opposing force, and that where it is influenced only by inertia, which is not a force at all, since it cannot cause motion. The use of the term resistance in both cases would seem already to have led to some confusion. Thus Maxwell—"Theory of Heat"—and Goodeve—"Principles of Mechanics"—both define work as being done against resistance, just as Rankine does; and it is only later on, and accidentally as it were, that the reader discovers that inertia is meant to be included as one of the forms of resistance in this definition.

112. In spite of the force of these arguments, it must be conceded, I fear, that the opposite practice to Rankine's has become so general, both here and abroad, that it is not probable it will again be modified. It seemed desirable to

bring out as clearly as possible the fact of the variation in question, and to put on record the grounds for wishing that the nomenclature originally introduced by Rankine had been retained. Having done so, I shall not attempt to combat any longer the prevailing fashion, and shall only aim at stating fully and clearly that which it lays down. Returning to the analysis in Art. 87, we shall agree to call Ps the Energy exerted, when viewed in reference to the moving force of A, and the Work done, when viewed in reference to the system B and C, on which the effect is produced. It may be asked why on this system we need retain the word work at all ; but, even supposing it possible to banish from the mechanical vocabulary so long-established a term, it is necessary to retain it, in order to keep in view the relations of cause and effect; as will be seen hereafter, when we come to treat of the conservation of energy. Accepting, then, the designation of Ps as the total work done, we have yet to fix names for the two parts into which it is divided, represented respectively by $-Qs$, and Bv^2. It seems desirable to have distinct appellations for these two parts of the total work done; and the clearest would, in my opinion, be "work of position" and "work of motion" respectively; the former being expended in altering the position of A with respect to another centre, namely, C, while the latter is expended in increasing its velocity. But, bowing as before to general custom, I shall designate them as the potential work and the kinetic work, since, as we shall shortly see, they are properly correlative to the potential energy and the kinetic energy respectively.

113. *Principle of Equivalence of Work and Vis Viva.*— There is yet another caution which needs to be given before leaving the question of nomenclature. Mention is often made of the principle of the equivalence of work and *vis viva*, and this is stated to be that the work done on the system is equivalent to the change in the *vis viva*. Now as we have defined the work done to be represented by Ps, it would follow from the principle, as thus stated, that Ps is equivalent to Bv^2, whereas in reality it is $(P - Q)s$ which is its equivalent. What is meant by the work in this case is, therefore, the net difference between the total work done by

the effort and the work done upon the resistance; in other words, it is the kinetic work only, as above defined. This fact should always be made clear when the principle is used as stated. In this, its correct form, the principle has, of course, already been proved in article 95.

114. *Conservation of Energy.*—We have now seen that, in the simple case under consideration, the following changes have taken place. There has been an exertion of energy on the part of the body A acting upon B, which is represented by Ps, and which may be looked upon as the action of a cause. Again, the effect due to this cause, or the work done, is divided into two parts:—(a) the potential work represented by $-Qs$, which means that the force Q of the centre C has been overcome through the distance s; (b) the kinetic work represented by Bv^2, which means that the velocity of the centre B has been increased from O to v. Now, by the general principle of conservation—namely, that effects live— we should expect that these two effects would be capable in themselves of acting as causes, to produce equivalent effects; and we have to see whether this is the case.

115. Let us first take the potential work Qs. In order to examine this effect by itself, let us suppose that, when B has traversed the space s, the velocity v and the force P are both annihilated, so that B is left at rest, exposed solely to the attraction of C, which, as before, we may assume to act by a series of impulses. It follows that A will begin to move backwards towards C; and since the action of C, by our definition of matter, is independent of time, and is always the same at the same distance, it follows that the first impulse given to B, on its returning road, will be exactly the same in magnitude and direction as the last impulse which it received on its outward journey. Similarly the second impulse of the new set will be precisely the same as the last but one of the old set, and so on throughout; so that when B has reached the point from which it started there will have been an exertion of energy on the part of C, which is represented exactly by Qs, and which is, therefore, precisely the same in amount as the energy which was exerted by C during the previous movement, and was neutralised by the greater

force of A. At the end of this time B will have a velocity V, which is such that $B V^2 = Qs$.

116. We have here spoken of C as remaining fixed and B moving towards it, because that was the assumption with which we set out. But of course we may just as well assume B to be fixed and to draw C towards it by the equal mutual attraction subsisting between them; or—which would really be the case if B and C were left to themselves—that they both move towards each other under the influence of that mutual attraction. The only difference will be, in that case, that the distance s, instead of being measured along B's path only, will be measured partly along B's and partly along C's; being, in fact, in all cases, the distance by which the two centres have approached each other during the motion.

117. We may therefore say that on our definition of matter the potential work done upon B in the course of its motion renders possible the exertion of a precisely equivalent amount of energy, due to the mutual attractions between B and C.

118. Let us now consider the kinetic work Bv^2. To examine this question by itself, let us suppose that at the end of the space s the centres A and C are replaced by a single centre at C, acting with a force $(P - Q)$, that is equal in amount, but opposite in direction, to the net force which has generated the kinetic work Bv^2. As before, we may suppose this force $P - Q$ to act by impulses at intervals $d s$. Then, by the Second Law of Motion, each of these impulses will produce its full effect upon B, irrespective of the fact of B's present motion; it will therefore destroy a portion of B's kinetic energy precisely equivalent to that which was generated by any one of the n equal impulses which acted on B during its motion along the space s. Hence, by the time that n of these equal impulses, due to $P - Q$, have acted upon B, the whole of its kinetic energy will have disappeared, and it will be at rest. But in the meantime B has overcome the force $P - Q$ through the distance s, precisely as the force Q was overcome through the distance s in the former motion; and therefore, as explained in the last paragraph, the kinetic energy Bv^2 in disappearing must have generated an amount of potential energy, due to the

mutual attraction between A and C, which is represented by
$(P — Q) s$. This is therefore precisely equal in amount to
the energy by which the velocity v was originally generated
in B.

119. Here, as before, for the sake of clearness, we have
represented the energy as being destroyed by precisely the
same steps as those by which it was generated; but if we
only grasp the principle that the kinetic energy $B v^2$ fully
represents the effect of the energy originally expended upon
B by $P - Q$, it will be evident that this representation is in
no wise essential to the proof. We may suppose, for
instance, that A and C are both annihilated, and that B flies
on in a straight line, with undiminished velocity v, until it
comes into the range of another centre D, whose force R
may be a repulsive one. The centre B will then be
gradually stopped, and will come to rest in a distance S,
which will depend on the value of R, but which will certainly be such that $R S = B v^2$; inasmuch as R S is known
to represent the energy which R will, during the passage
over the space S, have exerted on B, and this energy has
been expended in destroying the whole of the kinetic energy
represented by $B v^2$.

120. We may, therefore, say that, on our definition of
matter, the kinetic work done upon B in the course of its
motion renders possible the exertion of a precisely equivalent
amount of energy, due to the mutual action between B and
any other centre within whose range it may come.

121. Let us now gather our results together. We started
with an amount $P s$ of energy exerted by A. We saw that
the effect of this exertion was the performance of work, but
work under two different forms, namely, potential work,
represented by $Q s$, and kinetic work, represented by $B v^2$.
We then found that the performance of each of these
amounts of work rendered possible the exertion of a fresh
amount of energy, not due like the first to the action of A,
but precisely equivalent in amount to the original energy
exerted by A in the two cases. In other words, the kinetic
work $B v^2$, done by A upon B, gave B the power of subsequently doing the potential work represented by $(P - Q) s$
or R S respectively, in the two cases described in Art. 118,

119; and the potential work Q *s*, done by A upon B, gave B the power of subsequently doing the kinetic work represented by B V² (Art. 115), which can, of course, by a further operation, be converted also into potential work of equivalent amount. Hence, if we define Energy, which we have not yet defined, as the power of generating potential work, we see that the energy of A with regard to B has, indeed, been reduced in the original action by the quantity P *s*, since the impulses which have gone to do that work have had their effect, and cannot, by any action of A and B, be again renewed; but that the energy of B, or its power of doing work upon other centres, has been augmented by precisely the same amount. Hence, we see that, taking the system as a whole, there has been no gain or loss of energy during the action. This is the principle of the Conservation of Energy as applied to this particular case.

122. It will now be advisable to recur to the assumptions (Art. 94) with which we started on this investigation, and see how far they affect the generality of the principle we have just stated. In the first place we assumed that A is fixed. Since in every case of mechanics it is necessary to assume some fixed point, and to consider the motions relatively thereto, there is no difficulty in making A that point. In practice the centre of the earth may be considered as fixed for all questions of terrestrial mechanics, and the centre of the sun as fixed for the purposes of astronomy.

123. Again, we assumed that C is fixed; but if C have a motion in the direction C A, the only effect will be that we shall have to diminish the quantity Q*s*, expressing the energy imparted by C, by the quantity Q*s*¹; where *s*¹ is the space described by C, during the time of the motion in the direction C A. The effect will therefore be the same as if C were at rest, while the amount of Q was diminished in the ratio $s - s^1 : s$. There would thus be a diminution in the potential work, and, of course, a corresponding increase in the kinetic work. If C have a motion in the opposite direction, the potential work will be in like manner increased.

124. Further, we supposed C and A to have no mutual action. In reality this is not, of course, true, by our definition of matter; but in many cases C and A may be

fixed with regard to each other—as where coals are wound
up from a pit by a steam engine at the surface, which is
fixed with regard to the earth—and this amounts to the
same thing. Moreover, as we shall see hereafter, the forces
of cohesion are very great at insensible distances, but are
quite inappreciable at sensible distances; hence, in considering, for instance, the case of a rope in tension, we may
treat any section as being influenced by the two sections on
either side of it, but not by those beyond. If, however, A
does attract C, the effect is to move C in the direction of
A, and thereby make the distance between B and C, at
the end of the motion, less than in the former case.
Thus, let R be the force with which A acts on C, and
S the distance through which C has moved along the
line C A. Then A will have exerted the additional
energy R S, which will all take the form of kinetic work
done upon C. On the other hand, the energy P s
exerted on B will be just the same as before; but the
part of it which takes the form of potential work will be
reduced from Q s to Q $(s - S)$, because the number of
impulses distributed over the space S will not have been
given. Now the effect of this on B will be exactly the
same as if, C remaining fixed, the strength of each
impulse had been reduced in the proportion $s - S : s$;
for in that case the potential work would be represented by

$$Q \frac{s - S}{s} \times s = Q (s - S) \text{ as before.}$$ But if the resistance

be reduced from Q to $Q \frac{s - S}{s}$, then the unbalanced effort

will be increased from $P - Q$ to $P - Q \frac{s - S}{s}$; and the

kinetic work, due to this unbalanced effort, will be increased

from $(P - Q) s$ to $\left(P - Q \frac{s - S}{s} \right) s$, or $(P - Q) s + Q S$.

Thus the kinetic work will be increased by Q S, which is
exactly the amount, as shown above, by which the potential
work is diminished. Hence the assumption that A acts on C
does not introduce any gain or loss of energy on the whole;

the energy exerted on C takes the form of kinetic work, and the energy exerted on B partly of kinetic and partly of potential work as before, but divided in different proportions.

125. The assumptions still remaining to be considered are (1) that B is initially at rest; (2) that the forces P and Q are constant. Now with regard to the first, let us suppose that instead of being at rest the centre B has an initial velocity V in the direction B A. (If the velocity is in the opposite direction B C, the demonstration will be the same, simply writing — V for V throughout.) Then by virtue of this velocity it will also have kinetic energy represented by B V² (if B is half B's mass), which can be converted into potential work, as explained in Art. 108. Let t be the interval of time considered. Then, by Art. 69, since the net moving force (P—Q) has been acting on the mass 2 B during the time t, it will have generated in B—irrespective of B's initial motion—a velocity represented by $\dfrac{P-Q}{2\,B}\,t$, and will have caused B to describe a space represented by $\frac{1}{2}\,\dfrac{P-Q}{2\,B}\,t^2$. In addition to this B will have described in the same direction, by virtue of its initial velocity, a space V t. Hence the total energy exerted by A will now be represented by

$$P\left(\,V\,t + \tfrac{1}{2}\,\frac{P-Q}{2\,B}\,t^2\right);$$

and the total amount of energy which has to be accounted for at the end of the motion, is

$$B\,V^2 + P\left(\,V\,t + \tfrac{1}{2}\,\frac{P-Q}{2\,B}\,t^2\,\right)$$

Now the energy left at the end of the motion is as follows: (1) The potential energy, due to the potential work done in moving B through the space

$$\left(\,V\,t + \tfrac{1}{2}\,\frac{P-Q}{2\,B}\,t^2\,\right)$$

in opposition to the force Q. This is represented by

$$Q\left(V\,t + \tfrac{1}{2}\frac{P-Q}{2\,B}\,t^2\right)$$

(2) The kinetic energy, due to the final velocity of B, or to

$$\left(V + \frac{P-Q}{2\,B}\,t\right).$$

This is represented by

$$B\left(V + \frac{P-Q}{2\,B}\,t\right)^2 = B\,V^2 + V\,P\,t - V\,Q\,t + P\,\frac{(P-Q)}{4\,B}\,t^2$$
$$-Q\,\frac{(P-Q)}{4\,B}\,t^2$$

Adding the two expressions, we get for the energy left

$$B\,V^2 + V\,P\,t + P\,\frac{(P-Q)}{4\,B}\,t^2$$

This is exactly the same expression as that given above for the total energy which has to be accounted for. It appears, therefore, that there is again no loss of energy during the motion, and therefore the principle of the conservation of energy is not affected by the initial velocity of B.

126. Lastly we have the assumption that the forces P and Q are constant. This of course is never exactly true in the universe, although it is true within our limits of measurement in many cases, *e.g.*, that of a stone falling to the earth. But we may always assume it to be true for an indefinitely small interval of time. Hence for each such interval the conservation of energy will hold, and if so it must also hold for the sum of the intervals ; that is for any particular time that is considered. The energy exerted must of course be determined in this case by the methods of the integral calculus.

127. We have thus proved that the principle of the conservation of energy is true, with complete generality, for the case in which there are only three centres of force, A, B, C, lying in the same straight line. The extension from this to the general case of a free system of any kind, and with any number of centres, all of which act upon each other, we shall not give in detail, since it is to be found in any standard work on higher dynamics, and also comprises more analysis than comes within the scope of this treatise. The essential

features of the method are briefly as follows. By the principle of the composition of forces, we resolve all the forces acting on any given centre, and also the motion of that centre, into three directions, along three rectangular axes. By this means we reduce all the forces and motions to three straight lines, and consider these separately as in the simple case. Then, taking an indefinitely small interval of time, we class the forces which tend to move the centre in the direction of its actual motion as efforts, and those tending in the opposite direction as resistances. By the geometrical principle of the centre of position we can consider all the efforts as if they were a single effort, acting from a centre whose position and motion is known; and similarly we can consider all the resistances as a single resistance. The problem is then reduced to the simple case of three centres, in which the principle has already been proved; and the methods of the integral calculus enable us to combine the three equations found for the three axes into one general equation, which expresses, with the utmost generality, the principle of the Conservation of Energy.

128. Before taking leave of this much canvassed principle, it will be well to state it in its most general form, and recall briefly the definitions, &c., which it involves. The statement may be as follows:—"The energy of any isolated system of matter remains always constant, unaffected by the mutual actions of the forces which exist in the system."

129. Now, in this statement the following things must be borne in mind:

130. (a) The energy of the system means the power of doing work, or of overcoming force through distance—that and nothing else. It can be measured, like any other power, only when it has been exerted, and it is then measured by the amount of work it has done.

131. (b) The word force means force as defined in this treatise—that and nothing else; in other words, it means the cause of motion.

132. (c) The word matter means matter as defined in this treatise—that and nothing else; in other words, the system is a system of centres of force, acting upon each other by equal and opposite forces, which do not vary with the time,

and therefore are always the same when the distances apart are the same. These are the forces which are spoken of at the end of the statement.

133. (*d*) By an isolated system is meant one which is not acted upon by any forces from centres external to the system. Therefore, the principle is not strictly true of any body of matter less than the whole material universe, since the phenomena of light and gravitation show that every part of this universe is at least capable of being acted upon from every other part. There are, however, many cases where a system may for all practical purposes be treated as isolated, the actions of the rest of the universe being either allowed for or neglected.

134. It will be found that each one of the definitions just given are employed and needed at some point or other of the proof. If, in stating some proposition which is called the Conservation of Energy, the same terms are employed with any meanings at variance with the above, then that proposition does not express the principle of the Conservation of Energy, as it is quoted and applied by the great writers on mechanics throughout the world. This second proposition may, of course, itself be true, but it needs proof before it can be accepted, and certainly it cannot be accepted because the real principle of the Conservation of Energy has been proved.

135. *D'Alembert's Principle.*—The branch of mechanics called Rigid Dynamics is usually founded upon a principle known as that of D'Alembert, and generally laid down (Routh, "Rigid Dynamics," ch. ii.) as an independent deduction from the facts of nature, not to be proved by abstract reasoning. It will be well to show therefore—as has already been done by Thomson and Tait—that it is really a simple deduction from the elementary principles of mechanics, and in fact from the definitions laid down in this treatise.

136. The principle is expressed in words by saying that "The internal actions and reactions of any rigid system are always in equilibrium, and may be neglected in writing down the equations of motion." This, however, requires

some explanation. It is clear that the equation of motion of a point or centre which moves in a straight line, and is acted only by forces in that line, is

$$m \frac{d^2 s}{d t^2} = P - Q.$$

Here m is the mass of the point, P and Q are what we have called the effort and the resistance, $\frac{d^2 s}{d t^2}$ is the increment of the velocity, or the acceleration.

137. If, instead of one point, we have to do with a system of points, *i.e.*, with a body, as in rigid dynamics, the obvious way of forming the equations of motion would be to calculate the effort and the resistance for each individual point, and then write down its equation in the form just given. The general equation for the whole body would be formed by adding together all the particular equations thus obtained. Now in calculating the effort or resistance for any given point, there are two sets of forces to be dealt with—(1) The impressed forces, *i.e.*, those which act upon the system from without; and (2) the internal forces, or the forces acting between the points of the system themselves. Now the former set, the impressed forces, are generally few and simple, or may at least be regarded as such by neglecting those which are insignificant. But the latter set, the internal forces, will clearly, in any body of finite size, be immensely numerous and often obscure. Now D'Alembert's principle asserts that if all the equations were thus formed and added together, the whole of the internal forces would disappear from the resulting equation, being in fact in equilibrium amongst themselves, and therefore exercising no effect on the motion of the body. This being so, we need not stop to calculate these forces in the first instance (if, as in rigid dynamics, we are only concerned with the body as a whole), but can at once write down the resulting equation as follows:—

$$\Sigma m \frac{d^2 s}{d t^2} = \Sigma P - \Sigma Q,$$

where the symbol Σ signifies, as usual, that the sum of the quantity before which it stands is to be calculated for each point of the system.

138. This explanation relates to motion in one direction only. It can be generalised, in the way indicated in former cases of the same kind, by resolving the forces acting along three rectangular axes.

139. But this explanation is itself sufficient to show that D'Alembert's principle is only an extension of the Third Law of Motion, as the case is stated by Thomson; or, as we should here state it, a deduction from the definition of matter. For, by this definition, the forces which act between any two centres of force are equal in magnitude and opposite in direction; hence in any algebraical sum which includes both these forces, they will cancel each other, and the sum will be the same as if they did not exist. But it is evident that the algebraical sum of all the internal forces of any system will embrace both of the equal and opposite forces which act between any two centres of that system; hence in such a sum the whole of those internal forces will disappear. And this is precisely what D'Alembert's principle asserts.

140. *Elasticity.*—In our discussions upon energy we supposed throughout that the centres concerned were separated from each other by a finite distance, and that the forces acting between them were actually or approximately constant. But it is clearly conceivable that two centres may come within an indefinitely small distance of each other—in other words, may meet; and all experience shows that, as expressed in the definition of matter, the forces between the centres as actually existing are not constant, but vary with the distance. It becomes therefore necessary to consider what will happen if two centres meet; and to do this we must examine more closely what the laws of the forces acting between them really are. This branch of the subject, which is called Elasticity, involves entirely fresh considerations, and therefore requires some notice here.

141. Let us, as a preliminary, examine what would happen in the meeting of two equal centres, if the forces were really constant. Suppose them to start from rest, at a distance from each other of 2ft. Then by the principle of symmetry they would meet at the midway point, each with a finite velocity due to the action of the constant attractive

force over the space of 1ft. There being nothing to stop this velocity they would pass through each other—in this ideal state of things we need not discuss the question of penetrability—and go forward each in its old direction. But the attractive force on either centre, being now in the opposite direction to that of motion, would check this velocity, and destroy it at the end of the same space in which it was generated, *i.e.*, 1ft. Hence when each centre had arrived at the precise spot originally occupied by the other, it would be at rest. The circumstances would now be the same as at first; the centres would begin to approach each other again, would again pass through each other, and return to their original positions. This process would go on for ever, the two centres describing regular oscillations about the midway point.

142. Let us now make another supposition. Suppose that, when each of the centres had moved half way to the middle point, or through 6in., the constant attractive force was suddenly changed to an equal and opposite repulsive force. This would destroy the velocity thus acquired in exactly the same space in which it was generated. Consequently at the instant when the two particles met each other, they would both come to rest. The repulsive forces would then drive them asunder; but if, when the two were once more 1ft. apart, it was changed back into an attractive force, the velocities would be again checked, and the centres would come to rest precisely in their original positions. They would then again approach each other as before, and would thus continue to oscillate backwards and forwards, alternately approaching to and receding from the midway point

143. It is needless to say that nothing approaching either of these processes has ever been observed; and therefore we are justified in concluding that the forces of the universe are not constant. At the same time the two cases illustrate clearly a way in which a stable or conservative movement —or an oscillatory movement, using the word in its most general sense—may be produced by the action of attractive forces, or of attractive and repulsive forces combined; and we know that the world is, on the whole, in such a stable condition.

144. Let us now inquire what the laws of the forces acting in Nature can be made out, by observation and experiment, to be. On this head it must be confessed that our knowledge is as yet miserably imperfect. We have, indeed, one proved and almost perfect generalisation, due to the genius of Newton; and as it is the only one to which such epithets can be applied, we had better begin with it. It is generally stated as follows:—" Every particle in the universe attracts every other particle with a force varying directly as the product of their masses, and inversely as the square of the distance between them."

145. Using symbols, let Mm be the masses of the two particles, r the distance between them, and c a constant; then the moving force acting between the bodies is an attractive

force, represented by M$m \dfrac{c}{r^2}$.

146. I have called this generalisation "almost perfect," and for this reason. We know the character of the force, viz., that it is attractive; we know its law, viz., that it varies inversely as the square of the distance; we know its amount, at least as measured according to an arbitrary standard of mass. We do not, however, fully know its scope, viz., whether it extends to all matter that exists, or only to a part. Thus, it is known to hold in the heavenly bodies, by the case of double stars, but it is still disputed whether it holds across the void of space between those systems and our own. And if the principle of continuity be invoked to decide this in the affirmative, it would still remain doubtful whether it holds with respect to the luminiferous ether—a kind of matter whose existence Newton, of course, did not recognise. It would seem safe, therefore, to restrict the expression of the law of gravitation for the present to our own solar system.

147. Making this restriction, and remembering our definitions of a particle, as merely a small collection of centres of force, and also of mass, as expressing the number of centres comprised in a particle, we should formulate the law of gravitation as follows :—" Every centre of force in the solar system attracts every other centre with an equal force, varying inversely as the square

of the distance between them." By equal forces are, of course, meant forces which are equal at equal distances.

148. To give the proof of this law is no part of the present treatise. Assuming it to be true—as all competent judges assume—we have next to inquire whether it is the only law. In other words, whether it will, by itself, account for all the phenomena of the universe. It is obvious that this must be answered in the negative. Were two centres left to themselves under this law, they would rush together with a velocity which at the instant of meeting would become infinite, since, the distance being nothing, the force would then be infinite. What would happen I will not take upon myself to say, but at least it would not be anything like what we see around us. Nor is the case altered by the existence of other centres. A system starting from rest, under the action of gravitation alone, would coalesce in like manner at its centre of gravity. Hence there must be something beyond gravitation—something which acts as a repulsive force, and prevents the centres from thus dashing themselves against each other. Can we give the law, amount, &c., of this force as we can in the case of gravity? Unfortunately, we cannot. This is the great problem of physical science, which awaits the coming of a second Newton. But without being able fully, or even partially, to explain this force, we may at least glean a few facts respecting it. It must be practically insensible at sensible distances; otherwise, the law of gravity would not be found fully to account for the facts of astronomy and of falling bodies, which it is known to do. Hence it must diminish as the distance decreases; in other words, it must vary inversely as some power of the distance. But this power must be higher than the square; otherwise it would increase or diminish just as fast as gravity, and no faster, and would only have the effect of diminishing the apparent absolute value of gravity. These conditions are satisfied by assuming that the real law of force acting between two centres is represented, not

by the expression $m \, M \, \dfrac{c}{r^2}$, but by the fuller expression

$$= \mathrm{M}\left(\frac{c}{r^2} - \frac{c^2}{r^2 + n}\right),$$

where n is some positive quantity.[*]

149. It may perhaps be objected that the second term of this expression could not possibly disappear from view so completely as it does in all questions relating to the attraction of gravity at sensible distances. To examine this point, let us suppose that the attractive and repulsive forces are equal in amount at a distance of one-millionth of an inch, and also that the repulsive force varies as the fourth power of the distance. Then we have—

$$c \times (1,000,000)^2 = c^4 (1,000,000)^4$$
$$\text{or } c = c^4 (1,000,000)^2.$$

If the distance is one-thousandth of an inch, the expression becomes—

$$m \mathrm{M} \left[c (1000)^2 - c^4 (1000)^4 \right.$$
$$\text{or } m \mathrm{M} \left[c (1000)^2 - c (1000)^2 \times \frac{1}{1,000,000} \right].$$

Hence, at the distance of one-thousandth of an inch, the repulsive force will be only one-millionth part of the attractive force, and therefore quite insensible.

150. The assumptions here made are, of course, arbitrary; but they are sufficient to show how easily the repulsive force

---

* It will be noticed that in the text I have not introduced the conception of initial rotating motions in the centres. Sir Wm. Thomson appears to hold—see his recent lecture on "Elasticity as Possibly a Mode of Motion"—that the repulsive actions in nature may be accounted for by the fact that the centres composing each atom have—as on the vortex theory they would have—a very high velocity of rotation about each other, or about a common centre. Into the question of the adequacy of such motion to produce the observed facts I will not enter. If it should prove to be true, the repulsive action must still have the form of a force obeying something like the laws given in the text, just as that which is called centrifugal force, but is really an effect of motion, may be represented by a force whose law is expressed by

$\frac{v^2}{r}$. Meanwhile, it seems better to retain, for the purposes of explana-

tion, the simpler conception of an actual force; at least, until it is proved that we must replace it by the much more complicated and difficult conceptions of the effects of rotary motion.

may really exist at all distances, yet may be imperceptible by the most delicate measurements, unless at distances which are almost inconceivably small.

151. Assuming the law of force to be something like what has been described, let us now consider what will happen when two centres of force are left to its operation. We may consider the motion of one of them B, relatively to the other A, taken as fixed. Suppose, first, that B is placed exactly at the point of equilibrium, that is at the point where the attractive and repulsive forces balance each other. Then B will clearly remain at rest. Suppose next that B is slightly beyond this point of equilibrium. Then, the attractive force being slightly the larger, B will move towards A, and will pass the point of equilibrium with a certain small velocity. From this moment, however, the repulsive force will be the larger; the velocity of B will consequently be checked, and at a certain small distance within the point of equilibrium it will come to rest. The repulsive force being still the larger, the same operations will then begin in the reverse order; B will be repelled from A, and pass the point of equilibrium with the same velocity as before, but in the reverse direction, and will then be checked by the attracting force, and brought to rest exactly at the point from which it originally started. The same cycle will then begin again. In other words, B will continually describe, with regard to A, a series of small oscillations about the point of equilibrium. If B is placed at first slightly within this point, instead of beyond it, the same events will follow, but in the reverse order.

152. In either of the above cases, suppose a third force to act upon B, tending to move it toward A. Then, so soon as B is within the point of equilibrium, the repulsive force will be larger than the attractive force, and the excess will increase very rapidly as B continues to approach A. Hence this excess of the repulsive force will soon counterbalance the external force, and B will remain at rest at the new point of equilibrium thus defined, or rather will continue making small oscillations about it. If, on the contrary, the third force tends to move B away from A, then the attractive force will be in excess, will counterbalance the third force, and will form a new point of equilibrium further away from

A than the original one.   In the first case the force is compressive, and the net result is that, so long as the force acts, the distance between A and B will be permanently shortened.   In the second case the force is tensile, and the net result is that, so long as the force acts, the distance between A and B will be permanently lengthened.

153. Again, let us suppose that B starts from a point at a considerable distance beyond the point of equilibrium (or, which comes to the same thing, that it starts with a considerable impressed velocity towards A).   Then, by the time it reaches that point, the attractive force, which throughout this distance is largely in excess, will have imparted to B a very considerable velocity.   As soon as B has passed this point, its velocity will be checked by the excess of the repulsive force ; and it will be destroyed, and B brought to rest, somewhere in the very small space between the point of equilibrium and A, so that it will never come in actual contact with A.   For if A and B were in actual contact, the distance between them would be indefinitely small, and therefore the repulsive force would be indefinitely great. But the accelerating force which would destroy any given finite velocity v in any given finite distance s is simply given by the expression $\dfrac{v^2}{2s}$: and, however large v may be, or however small s may be, this will always have a finite value. Hence the effect will be that B will be stopped in an exceedingly short space and time—much too short for our measurement,—and will then have a very large excess of repulsive force acting upon it.   Hence it will begin to return with very great rapidity, will pass the point of equilibrium with the same high velocity, but in the reverse direction, and will then be checked by the excess of attractive force, finally coming to rest at the point from which it started.   If no other cause intervenes, the same cycle will then begin again.

154. Lastly, let us suppose that in the last case B becomes fixed in space at the moment when it is stopped by A, while A becomes free to move ; or, which comes to the same thing, let us consider the motion relatively to B, instead of rela

tively to A. Then, since A, by the Third Law of Motion, has exactly the same repulsive force acting upon it as B has, it will fly off with the same rapidity as was ascribed to B in the last section, will travel to exactly the same distance, and there will come to rest and begin to return, unless some other cause supervene.

155. It should perhaps be remarked that the point where A comes to rest will be very much further away from the point of equilibrium than the point from which A is supposed to have started. For when B is within the distance of the point of equilibrium from A, the excess of the repulsive over the attractive force increases, as B moves towards A, very much more rapidly than the excess of the attractive force increases, as A moves away beyond the point of equilibrium. This is due to the forces varying inversely with the distance. Thus let the point of equilibrium be at one-millionth of an inch, as before, and let the expression be $\left[\dfrac{c^1 (1,000,000)^2}{r^2} - \dfrac{c^1}{r^4}\right]$.

Then if B be half-a-millionth of an inch within this point, the value of the expression will be $c^1 (1,000,000)^4 (4 - 64)$, or $- 60\, c^1 (1,000,000)^4$. But if B be half-a-millionth of an inch beyond this point, the value of the expression will be $c^1 (1,000,000)^4 \left[(\tfrac{2}{3})^2 - (\tfrac{2}{3})^4\right]$, or $0\cdot 247\, c^1 (1,000,000)^2$, which is about $\tfrac{1}{240}$ of that given above.

156. I have preferred to place these deductions from the assumed forms of force together, because they follow naturally and clearly upon each other. It now remains to show how fully they accord with the facts of the universe, as relates to the behaviour of the particles of solid bodies in close contact with each other. Of course, the matter is greatly complicated by the fact that we can never observe the motions of single centres of force, or even single particles. What we observe are bodies, greater or less in size, but of which the adjacent particles act in various ways upon each other, and are also acted upon in various ways by external forces, such as gravity. Nevertheless, the effects here described are in many cases plainly discernible.

157. Thus, in accordance with Art. 151, the particles of any solid body do take up apparent positions of equilibrium

with each other—which, however, are known not to be really positions of rest, but centres of small oscillations which the molecules are continually describing. If an external force be brought to bear upon such a body, then, in accordance with Art. 152, the body becomes extended if the force be tensile, or shortened if the force be compressive; and, having thus taken up a new position of equilibrium, it retains it until the force is withdrawn.

158. Again, if a body be projected against another with considerable velocity, which is equivalent to the supposition of Art. 153, then, after apparently striking it, it flies back in the direction from whence it came, the reversal of the motion being effected far too rapidly for any ordinary means of observation to follow it. It is this property of rebounding which forms what is called the elasticity of bodies. Newton, who investigated it, found that the effects might be represented by supposing that, at the moment of impact, the momentum of the striking body was stopped by a very large repulsive force brought into existence by the action, and that, the bodies having thus been brought to rest, a force continued to act in the same direction, and drive the striking body back again towards the point whence it started. If it actually reaches that point, the body is called "perfectly elastic." As a matter of fact, no substance in nature is found to be perfectly elastic; some, however, as glass and ivory, approach the limit pretty closely, while others can scarcely be said to have any visible elasticity. For some time it was supposed that in practice the force of restitution, causing the rebound, was somehow less than the original force of impact, in a varying ratio, generally expressed by the letter e; and hence it was inferred that *vis viva*, or energy, was always lost in cases of collision. But it is now universally admitted that this difficulty simply arises from the fact that we can only observe the action of finite bodies as a whole, and not that of their minute parts. For instance, when a billiard ball strikes the cushion, it is but a very small area of each which is actually opposed to the other; all the rest of the billiard ball is caused to stop and to rebound by the lateral cohesive action of the parts nearer the centre. These actions set up movements between the particles, in

which more or less of the energy due to the impact becomes expended, and is not therefore available for the repulsion of the body as a whole. It is not doubted by anyone that the ultimate atoms of any body are perfectly elastic, as indeed, by the conservation of energy, it is necessary that they should be.

159. Lastly, if the body struck be free to move, instead of being fixed, the result stated in Art. 154 is actually seen to follow; that is, the struck body flies away with a velocity which depends on the momentum of the striking body, or, in other words, upon the force of the blow. Familiar instances are the striking of one billiard ball by another, the propulsion of a football, &c. In such cases the striking body may either be brought to rest, or may follow in the same direction as the struck body but with diminished speed, or may rebound again in the direction whence it came. These variations depend upon variations in the masses and velocities of the two bodies, and need not here be considered further.

160. It appears, therefore, that the fundamental facts of elasticity are all accounted for by the hypothesis that the law of the force subsisting between any two centres is substantially of the character represented by

$$\left(\frac{c}{r^2} - \frac{c_1}{r^{2+n}} \right).$$

Of course we do not affirm that this is its exact representation. It may be much more complicated; e.g., there may be other factors which may bring about the difference existing between the chemical elements considered as kinds of matter. But the fact remains that, as regards the general phenomena of mechanics, the type given above must be an approximation to the truth, and we may hope that the progress of physical science will ere long enable us to fix it more exactly.

161. I do not propose to continue these papers further; but I believe that all the main principles on which the science of mechanics is founded have been touched upon, and have been shown to flow directly from the definitions as to motion, force, and matter laid down at the beginning, and from the

two established generalisations, which I have called the principles of conservation and of symmetry. If so, I have achieved all which I proposed to myself at starting. I have only two supplementary remarks to make. One is that these papers will not have been useless if they have brought into light, and in some measure dissipated, what appears to be a somewhat common mistake, namely, that in the third law of motion, Action is always to be interpreted as Force, so that, whenever we have a force acting, there must always be an equal force opposed to it. Thus, when an engine is taking a train out of a station, and getting up its speed, it is held that the pull of the engine is no greater than the resistance of the train! It need hardly be stated that this idea derives no countenance from Newton—who, as Thomson and Tait have shown, saw that the action was measured in dynamics by the energy exerted—or from any other authority; that it is altogether at variance with experience; and that to admit it would simply annihilate the present science of dynamics. The other remark is merely intended to guard against any possible conception, that among the great writers on mechanics there is any fundamental difference of opinion as to the physical foundations of the science. It is just possible this might arise from the fact of my having taken occasion to point out the difference in terminology, as to the term Work in particular, between the works of Rankine and those of later writers; and I therefore repeat that the difference is one of terminology only, and in no way affects the fundamental principles or conceptions of the subject. Anyone who doubts this may be recommended to study the books in the annexed list, which have been consulted in connection with these papers, and are here tabulated for the convenience of students of the science.

Newton's Principia (Glasgow, 1871).

Rankine, Applied Mechanics (1st and 3rd editions).

Rankine, Miscellaneous Scientific Papers (Griffin, 1880).

Rankine, Rules and Tables (Griffin, 1875).

Maxwell, Theory of Heat, Edit. 1877 (Longmans).

Maxwell, Matter and Motion (S.P.C.K.).

Thomson and Tait, Elements of Natural Philosophy (Clarendon Press, 1873).

Navier, Application de la Mecanique, by St. Venant (Paris, 1864).

Reuleaux, Theoretische Kinematik (Brunswick, 1875).

Clausius, Mechanical Theory of Heat (Macmillan, 1879).

Balfour Stewart, Lessons in Elementary Physics (1873).

Whewell, Mechanics (Cambridge, 1819).

Goodwin, Course of Mathematics (Deighton and Bell, 1857).

O. Byrne, Practical Mechanics (Spon, 1872).

Twisden, Practical Mechanics (Longmans).

Routh, Rigid Dynamics (Macmillan).

Tait and Steele, Dynamics of a Particle (Macmillan).

Herbert Spencer, First Principles (Williams and Norgate, 1867).

Todhunter, Mechanics for Beginners (Macmillan, 1878).

Moseley, Mechanical Principles of Engineering and Architecture (Longmans, 1843).

Goodeve, Principles of Mechanics (Longmans, 1880).

Magnus, Lessons in Elementary Mechanics (Longmans, 1881).

LONDON: PRINTED BY G. REVEIRS, GRAYSTOKE-PLACE, FETTER-LANE.